Electronic Business Revolution

Springer

Berlin
Heidelberg
New York
Barcelona
Hong Kong
London
Milan
Paris
Singapore
Tokyo

Peter Cunningham is President of INPUT, an Electronic Business information and marketing services company. He analyzes electronic business for companies in the US, Europe and Japan.

Friedrich Fröschl is President and CEO of Siemens Business Services (SBS), a leading European IT services company with more than 20,000 employees devoted to developing and operating electronic business.

Peter Cunningham · Friedrich Fröschl

Electronic Business Revolution

Opportunities and Challenges
in the 21st Century

With 64 Figures

Springer

Peter Cunningham
INPUT
Washington D.C.
1921 Gallows Road
Vienna, VA 22181-3900, USA
pac@input.com, www.input.com

Dr. Friedrich Fröschl
Siemens Business Services
Otto-Hahn-Ring 6
D-81739 Munich, Germany
friedrich.froeschl@mch20.sbs.de, www.sbs.siemens.de

This book was written, edited and produced through the collaboration of a geographically distributed international team that—besides the authors and Springer-Verlag—consisted of the following persons: Mark Dietrich, Axel Grueter, Geoff Nairn, Edward Rosenfeld and Persia Walker. Yukom Verlag (www.yukom.de) was responsible for editing and preproduction of all text and graphics.

ISBN 3-540-66211-1 Springer-Verlag Berlin Heidelberg New York

Library of Congress Cataloging-in-Publication Data applied for
Die Deutsche Bibliothek – CIP-Einheitsaufnahme
Cunningham, Peter: Electronic business revolution: opportunities and challenges in the 21st century / Peter Cunningham; Friedrich Fröschl. – Berlin; Heidelberg; New York; Barcelona; Hong Kong; London; Milan; Paris; Singapore; Tokyo: Springer, 1999
ISBN 3-540-66211-1

© Springer-Verlag Berlin Heidelberg 1999
Printed in Germany

Cover design: Künkel + Lopka, Heidelberg
Typesetting: Yukom Verlag
SPIN: 10751140   45/3012   5 4 3 2 1 – Printed on acid-free paper

Dedication

This book is dedicated to Dr. Volker Jung. As a member of the Board of Siemens AG and mentor of the company's largest business area, Information and Communications, he has assumed a pioneering role in the IT and telecommunications industry. Under his leadership, Siemens bundled its comprehensive portfolio under one organizational umbrella in order to drive the convergence of information and communications technologies—the prerequisite for the Electronic Business Revolution that is the subject of this book.

Table of Contents

Introduction

This book is about a revolution. Electronic Business will create changes similar to those brought about by the Industrial Revolution. It will impact our business, society and governance.

It is happening now—not in 10 years or even 10 months, but now! Historians will look back at this period and identify it as the time the revolution started.

What is it? Electronic Business is the integration of IT and particularly the Internet into business processes to change organizations and create new ones. It particularly affects the interfaces between and among organizational entities and units. In the past 30 years, IT has played an important part in making processes more effective and efficient but it has not changed them. Electronic Business changes them.

In Electronic Business, IT becomes embedded in business processes as opposed to the prior model, where IT was a support function. A critical difference is whether or not the process can operate at all if the IT system is inoperable. A reservation clerk that uses a reservation system to make airline reservations is not an example of Electronic Business. An airline system where a passenger makes a booking directly, gets an electronic ticket and walks onto the plane after having been identified by a retina scan is an example of Electronic Business.

The Internet is the catalyst. It will have an impact like other great communications developments: the telephone, cars/highways, railroads, etc. Its impact will be as great as or greater than that of the printing press.

There are threats as well as opportunities. Europe and its companies have some enormous problems from Electronic Business to deal with. Our purpose is to point out the challenges.

This book is about the "new," the future rather than the present. It is not a review of the literature or current news. We are looking for things that are possible and have not yet happened, or are just happening now. The examples we present are selected as pointers to the future or possible futures. We emphasize forecasts and predictions.

Our intent is to write for the thoughtful executive in private or public life; it is not a tutorial on technology, a "how-to" book or an academic treatise full of references and caveats. It provides a viewpoint that executives can use as the basis for their actions in analyzing and planning their organization's Electronic Business future.

We address the question "Why?": Why is this happening? Why is it happening now? Why is it happening in some places/industries and not others, etc.?

This book is based on our many years of experience of planning for and applying IT to some of the world's leading organizations. This includes research into computers, networks, software, services, electronic commerce and EDI, vertical industry use of IT, etc.

"Revolutions" Do Occur

Electronic Business is a revolution. It is not business as usual; it is not conducting old business in a new way. It is a radical departure from the past.

In a revolution, kings lose their heads, leaders lose their power and unknowns become the leaders. The last revolution in IT was the microcomputer or PC revolution. Microsoft, Cisco and Intel profited dramatically, while Wang, CPT, Burroughs, Univac, CDC, Data General and others did not. They survive in some cases, but not as the stars of the industry they once were.

The Electronic Business revolution will change the rules of the game, just as other revolutions did:

In transportation, the airplane and the car supplanted the railroads and ships for passenger traffic. Railroads themselves had largely replaced canal and coach traffic, and steam ships displaced sailing vessels.

Planes, trains and steamers generally moved the same commodities faster and cheaper. Other revolutions changed what was moved. For example, containers revolutionized shipping. As a result, an entire industry built around the London docks disappeared.

The emergence of fast-food franchises such as McDonald's and super retailers such as Wal-Mart was an unforeseen consequence of a transportation revolution—in this case, the US interstate highway system.

At a time when traditional retail chains such as Sears, J.C. Penney and Woolworth's were still investing in downtown areas, Sam Walton was building his first discount stores in green fields along the new interstate highways. People thought he was mad, but Sam recognized that the interstates and cars would bring customers to his stores. Being next to the freeways made delivery easy. Being out of town meant that space was cheap and plentiful, so he could have a great variety of goods and "pile 'em high, sell 'em cheap!" Now, Wal-Mart is the world's largest retailer. Does this "resonate" with the analogy of the Internet to the interstate highway system? Who will be the Sam Walton of the Internet Superhighway? Will it be Jeff Bezos of Amazon.com?

Another example is how the electric power network revolutionized manufacturing. Before the network, work had to be brought to the power; now, networks deliver the power to work.

At the end of the 19th century, people speculated as to which would be the largest city on Earth: Would it be Buffalo, which is next to Niagara Falls, the greatest source of hydro-power? Or might Liverpool, atop the world's (then) largest-known coal deposits, prevail? Both cities were easily accessible by sea. Then along came electric power networks, which obviated the special advantages that location had given Buffalo and Liverpool. The adoption of electric power networks was dependent on numerous fights over standards: AC or DC, what voltage, what cycles, just to mention some of the controversies. These were very similar in many ways to the recent fights over telecommunications network standards.

Just as power grids changed manufacturing, the Internet is changing business. Before its emergence, proprietary communications protocols and standards barred Electronic Business. Every national and business interest fought to protect its own IT (computer/communications) industry. Such squabbling has been eradicated. The Internet (TCP/IP) is the standard and it has a standard international "plug" (the browser), unlike electric power and telephone systems, which have plugs that differ according to the national company (sometimes several within a country) that is providing the service.

Parenthetically, the step-down motor was the 19th century's equivalent of the microcomputer. Both enabled work to be distributed in a different manner; essentially, the matching of power to the job. Little jobs could be done with small machines; big jobs with big machines. No longer was it necessary to use a sledgehammer to crack nuts.

Developments in communications have precipitated most of the great revolutions in human history. Arguably, democracy has thrived in the 20th century because of the rapid spread and widespread availability of communications media: newspapers, radios, televisions, faxes and finally the Internet. Political decision-makers recognize the importance of these information channels: Dictatorships seek to control them and NATO bombs television stations as "military targets."

Computer communications networks promise to revolutionize business, social and government processes in the 21st century. They will deliver IT power to worksites. They will enable and encourage new work distribution patterns that differ significantly from those of the 20th century, with its acres of office space piled atop acres of office space.

The elevator made multi-story buildings and offices possible, but their height was limited until the telephone was invented and replaced messengers. The telephone makes the World Trade buildings usable. The Electronic Business revolution may make these buildings obsolete.

The threat to any company, country or geographic entity is obvious: If it does not have the network or if it has one that cannot compete in terms of performance or price, it will lose competitiveness fast. Work will move elsewhere.

The Internet in the 21st century attacks geography. Because most of our institutions are geographic in nature, the implications are sweeping. However, the Internet will not kill culture. It actually enhances it, enabling you to be Italian in Sydney, Chinese in Rome, Australian anywhere. This is "walkabout" on a global scale.

Europe is well-positioned to take advantage of the Internet, provided it has the individual and institutional fortitude—"the guts"—to make the changes necessary to lead rather than follow. Its governments must let go and encourage the revolution to develop along its own lines. No one knows for sure where the revolution will lead. Indeed, few people even have a vague idea. That makes directed management of this process impossible. Nevertheless, the US government has set an example with its strongly open, supportive attitude since 1995 (notwithstanding its encryption and pornography missteps), when it launched the "Information Superhighway" and "Electronic Government" (EG) initiatives. (It is interesting to note that electronic government was debated in the USA before Electronic Business).

What Is "Electronic Business"?

Electronic Business is the execution by electronic means of interactive, inter-organizational processes. Here "electronic" is a combination of telecommunications and computing capabilities.

Electronic Business forms an umbrella for a series of distinct electronic processes along the chain from supplier to consumer, most of which have physical analogs. These Electronic Business processes in fact encompass the entire spectrum of human activity: from commerce to finance, from education to entertainment and from government to religion:

• Electronic Commerce is the business-to-business (BTB) buying and selling of physical goods (Trade).
  - EDI Commerce is the traditional, hard-wired, predetermined trade common in the automotive, aerospace and large retailer environments for primary goods (those needed to produce the product).
  - Internet Commerce is the new, open trade that is penetrating all industry segments. It is particularly used for secondary goods (those that are ancillary to the production of the product; office supplies, for example).

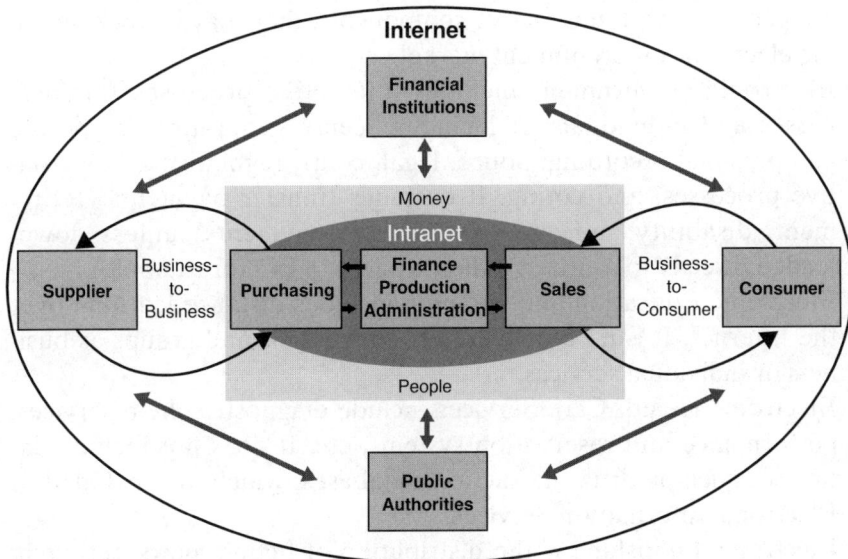

Figure 1. The Electronic Business framework

- Electronic Retailing is the business-to-consumer (BTC or CTB) buying and selling of physical goods. This is often referred to as BTC electronic commerce. It includes auction and reverse auction sales.
- Electronic Banking is the electronic buying and selling of banking products and services for businesses and individuals. It does not include settlement; the value of settlement is included in the other Electronic Business market sectors.
- Electronic Finance and Investment is the electronic trading of stocks, bonds and other financial instruments by businesses and individuals.
- Electronic Insurance is the buying and selling of insurance policies for business and individuals. It also includes the investment, claims submission and payout processes.
- Electronic Travel Services include the purchase of products and services such as tickets and reservations by individuals and businesses.
- Electronic Entertainment and Recreation includes the plethora of games, gambling, hobbies, recreational and interest activities available on the Internet. The interactive criterion means that traditional, one-way broadcasting is not included. But the Internet can be just another medium, like CDs and radio waves, for the transmission of "software" such as movies, music, sporting events and shows. Since these are available on demand by the buyer and/or are packaged with interactive components, they may be included in the electronic entertainment market.
- Electronic Government includes interactive processes for businesses and individuals. It includes license generation; corporate and personal recording; police, legal, court, regulative and legislative processes; and voting. It excludes transfer payments (retirement, disability and support program payments) unless downloaded directly to a citizen's client system (a PC, for example).
- Electronic Education and Training includes distance learning over the Internet. It can be delivered to individuals and groups as business or individual services.
- Electronic Health Care Services include diagnostics, help services, performance and reservation systems, etc. It does not include databases such as drug interaction databases, which are included in Electronic Information Services.
- Electronic Publishing is the distribution of fiction, news, research, special interest and other information previously only available in

books, magazines and newspapers. This includes the creation of new forms of collaborative works.

- Electronic Information Services is the buying and selling of individual and business information; it is composed primarily of databases that are searchable by products such as Alta Vista, Yahoo and the like. It includes credit, financial, personal (including employment, genealogy and DNA) and investment information. It includes portal and hub services.
- Electronic Community Services includes sporting, religious, trade union, club, association and other groups; it includes "chat" and other services that involve communities of businesses and individuals and that are not included in other Electronic Business segments.
- Electronic "Other Services" includes the buying and selling of services not otherwise categorized. These include construction services, utility services (such as electricity), telephone services, accounting and legal services, staffing services (only the hiring fee is included in the market rather than the compensation/billings of the individuals thus employed/contracted), translation services, etc.

There are also internal Electronic Business systems that provide electronic marketing, customer service, recruiting, training and more. These systems (Enterprise Applications Solutions) are only discussed in this book to the extent they are enablers of Electronic Business.

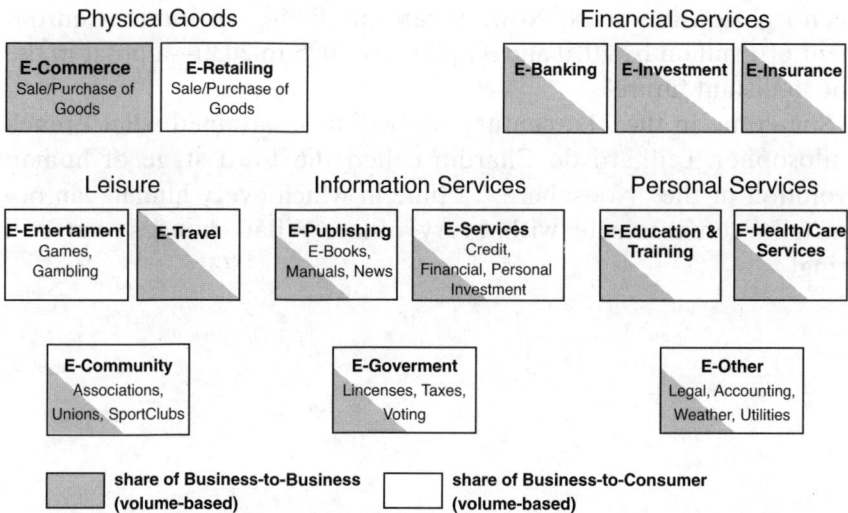

Physical Goods

E-Commerce Sale/Purchase of Goods	E-Retailing Sale/Purchase of Goods

Financial Services

E-Banking	E-Investment	E-Insurance

Leisure

E-Entertainment Games, Gambling	E-Travel

Information Services

E-Publishing E-Books, Manuals, News	E-Services Credit, Financial, Personal Investment

Personal Services

E-Education & Training	E-Health/Care Services

E-Community Associations, Unions, SportClubs	E-Goverment Lincenses, Taxes, Voting	E-Other Legal, Accounting, Weather, Utilities

share of Business-to-Business (volume-based)   share of Business-to-Consumer (volume-based)

Figure 2. Electronic Business structure

Some people put all the above Electronic Business activities under e-commerce and that is a disservice to the community. Each of the above processes (note we use "processes" not "industries") has unique characteristics and communities that warrant separate treatment. Electronic commerce, BTB trade in goods, by itself will be a multi-trillion dollar or euro component in the early 21st century; it deserves to be treated separately.

Electronic Business is about trade. Today, about 50 percent of manufactured goods are traded internationally. For services, this proportion is much lower: around 10 percent to 15 percent. We expect that Electronic Business will increase the proportion of international trade, particularly in services.

In this book, we consider Electronic Business from the perspective of the user rather than that of the vendor. We examine what makes Electronic Business possible: the "Enablers." This term refers primarily, but not exclusively, to the Internet and the Web. We will also touch on support and other necessary services.

This book is addressed to business leaders across all industrial and governmental boundaries. It is meant to be a guide to the opportunities of Electronic Business and an aid to surviving the seismic changes of the future. That's why the focus of the book is on the applications of Electronic Business.

In 1994, INPUT forecast that there would be 200 million people worldwide connected to the Internet by 2000 and that the Internet-related industry would be $200 billion. Our estimates appear to have been right on the mark. Now, we are predicting an Internet enrollment of 1 billion by 2010 and 3 billion by 2025 to 2030—a point in the not-so-distant future!

Sometime in the 21st century, we will have attained what French philosopher Teilhard de Chardin called the third stage of human evolution or the "Noosphere," a time in which every human can potentially communicate with every other. What changes that will bring!

1 Whirlwind of Change

The Internet is bringing a whirlwind of change to the business world and enabling new and diverse ways of trading that are collectively known as Electronic Business. To compete in this emerging digital economy, European businesses large and small will need to fundamentally transform business models, change the way they work and think, and form new relationships with their trading partners and customers.

Electronic Business has only just arrived. It is developing more quickly in some regions and industries than in others. As yet, it accounts for only a very small proportion of business even in those sectors where it is growing fastest. But it is the rate of expansion that is important; in many areas it is growing almost exponentially. This new business philosophy is here to stay and it is evolving rapidly, creating a wave of change for all businesses and organizations.

The IT industry is riding this wave by applying the Internet to its own business models. Companies such as Dell and Cisco have achieved success on the back of the Internet boom and are no longer viewed as simple hardware vendors but as full-fledged members of the new digital economy.

IBM is transforming itself into a major Electronic Business; the value of its Electronic Business transactions has grown from $1 billion to $11 billion in just three years. Siemens Business Services, one of Europe's largest IT services providers, is also heavily involved in this new Internet industry. It has developed a strategy, called "e-SPEED," that is designed to exploit the convergence of information and communications technologies and apply it to develop Electronic Business solutions.

Although slow to start, the telecommunications giants are rapidly moving into the field. Europe's "power players" such as Deutsche Telekom, France Telecom and BT, as well as US giants such as MCI/Worldcom, AT&T and SBC, recognize that while the Internet phenomenon threatens their very existence, it also provides them with unparalleled opportunities.

Besides these established players from the mainstream IT and telecommunications industries, the Internet is creating its own heroes—fast-growing young companies whose whole business philosophy is based on the Internet. One of the most famous examples is undoubtedly Amazon.com, the online bookstore.

But there are many more, including E*Trade, an online share trading company; America Online, the big Internet service provider that is also at home in Europe; and a host of other smaller, strange-sounding companies such as @Home Networks, eBay, Lycos, Real-Networks, Cyberian Outpost and Inktomi.

Their names are unknown to many, their profit often nonexistent and their business models unproved, but they have attracted intense interest, particularly from investors looking for the next Amazon.com.

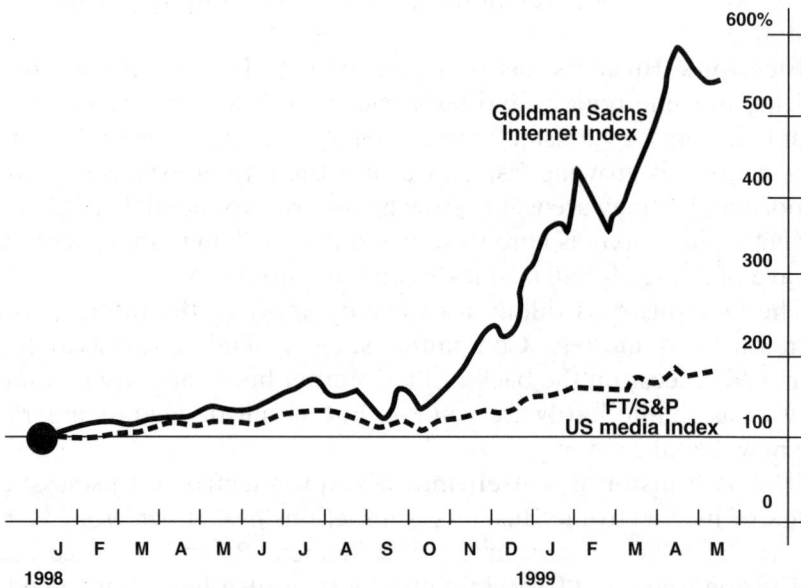

Figure 1. A comparison from January 1998 through May 1999 of Internet share prices based on the Goldman Sachs Internet Index with those of traditional media companies based on the Financial Times/Standard & Poor's US Media Index shows above-average growth rates for the Internet companies (January 1, 1998=100)

Most Internet technologies have been developed in the United States, but Europe is waking up to Electronic Business and starting to create home-grown Internet players. In fact, some of the premier companies driving the Electronic Business trend are European. One of Europe's pioneers is Intershop, an e-commerce software house, created in the early 1990s by a group of programmers from eastern Germany.

Surf onto the popular AltaVista Web site, owned by Compaq Computer. Click on the bottom of the page and you will find that AltaVista's query refinement technology was developed by the École des Mines in Paris.

Electronic Business is not just something that affects the IT industry alone. On the contrary, the Internet will touch and transform almost every industrial sector and commercial activity, creating new opportunities for those companies that can spot the trends—and serious problems for those left behind.

This chapter seeks to explain what Electronic Business means today and how it is unfolding as the competitive factor that could make or break your company.

Inevitably, the choice of cases is subjective and limited to those operating today. Nevertheless, our selection is designed to be reasonably comprehensive and, most importantly, to demonstrate that Electronic Business is not just happening in the United States, but is indeed a phenomenon sweeping both Europe and the world.

Why Me, Why Now?

It's not just about technology, and it's not just about the future. Electronic Business is happening now and those who are making it happen realize that while Internet technologies are essential to doing business in the digital age, they are not by themselves sufficient.

There are a host of issues to consider. Some are specific to a business, sector or country; others are more general challenges. Perhaps, the biggest barrier to the adoption of Electronic Business, particularly in Europe, is the perception that Electronic Business is something for big companies and exotic start-ups.

Silicon Valley may seem a long way, both geographically and culturally, from the valleys of the Rhine, Rhone or Po, but the technologies developed in California can be applied equally well in a European context, as many European companies have discovered.

Here are just a few examples, chosen at random, of how Electronic Business is transforming Europe's economy:

• **Kingfisher**, a UK home improvements group, plans to open a Web site in France to sell goods over the Internet. (But why open it in

France? In the Internet world, simply opening a French Web site is sufficient.)

- **Swiss Railways** runs a Web site that allows travelers anywhere in the world to not only look up timetables but also buy tickets on-line.
- **Osram**, a leading European lamp manufacturer, has developed an Internet-based information system for its sales department to improve customer service. The return on investment is more than 200 percent.
- **Sweden Post's** e-commerce service, **Torget**, gives access to more than 100 online providers and receives an average of 2.5 million "hits" per day. (A "hit" is one person accessing one Web site through the Internet.)
- **Computer Solutions and Finance**, a small UK computer reseller, lets its business customers buy goods via an online catalog accessed through the Internet.
- **Tabacalera**, Spain's leading tobacco company, and Cortefiel, a High Street fashion chain, have created Via Plus to sell products to consumers via in-store Internet kiosks.
- **RAS**, a big Italian insurance group, sells motor insurance policies online via a new direct sales unit, Lloyd 1885. The online rates are 15 percent cheaper than those bought over the phone.
- **Charles Stanley**, a traditional London stockbroker, has created Xest, an execution-only share trading service on the Internet that is authorized throughout the European Union.
- **3 Suisses**, a French retailer, attracted 20,000 users a day to its World Cup '98 merchandising site where soccer fans from around the world could choose from a selection of 400 items.

The only common denominator connecting these organizations is an appreciation of Electronic Business as a powerful new force that can transform their businesses. It can streamline their distribution channels, link suppliers, extend their product range and expand their customer base. In some cases, it has changed not only how individual businesses function and interact, but also rewritten long established rules and transformed the corporate culture.

New Rules in an Ancient Game

The Internet may seem to be a unique phenomenon, but this is not the first time that technology has dramatically transformed trade and commerce. The keeled-hull ships of the Phoenicians in 2000 BC made it possible to sail against the winds and go beyond the shores to the high seas of the Mediterranean. By doing something never before possible, the Phoenicians broke the limitations imposed by geography and developed into a flourishing trading nation.

Technology is important, but by itself it is not sufficient to transform an economy. The full potential of a technology always takes years to develop. Take the development of electrical power, for example. British physicist Michael Faraday is credited with producing the first electrical generator in 1831. Without commercially available generators and an efficient distribution network, however, electricity was initially little more than a scientific curiosity. Electricity's potential to transform our lives took 50 years to develop.

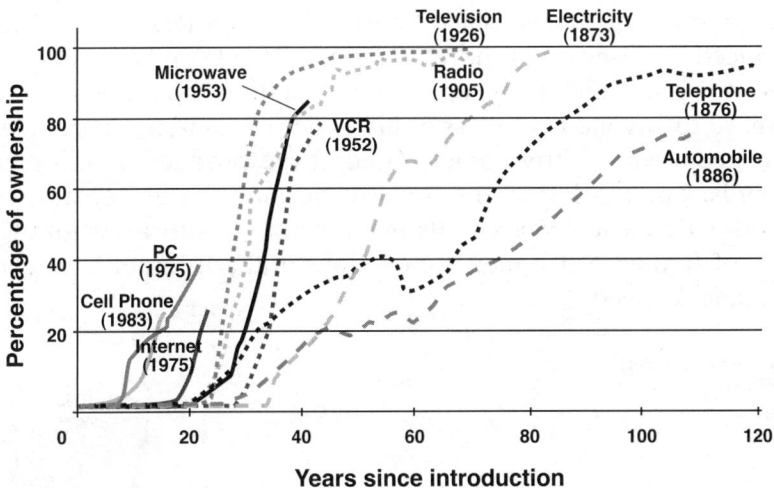

Figure 2. Compared to other technologies, the Internet has just started (DARPA report 1998)

One turning point occurred in 1881 when Thomas Edison, a US inventor, built a generator and started supplying electricity to more than 80 customers. Two years earlier, Edison had invented the light bulb and the early users of electricity were mostly interested in lighting. Factories also were quick to use the new electricity for lighting, but slow to use it as primary energy source. In fact, Edison's first business venture was not a success because it failed to recognize the importance of the networks.

Not until the networks and the stepped-down motor made their appearance did electricity achieve its breakthrough. The network took the power to the work and the stepped-down motor allowed power to be applied to each task at the optimum level. Factories, mines and other institutions began to replace their old power systems and build new factories based around the new systems. In the first decade of the 20th century, fundamental changes in production occurred. Factory structures were streamlined and key processes, such as materials handling and manufacturing flows, were made more efficient.

In Electronic Business, the principles of the packet-switching data networking technology that underpin today's Internet were demonstrated in 1966, while the first demonstration of an e-mail system took place in 1972.

It is only in the past five years that these technologies have been widely adopted and promoted by the commercial IT industry, ushering in today's digital age and an explosion in Internet users.

But the pace of change is accelerating. Tim Berners Lee at CERN introduced the concept of the World Wide Web (WWW) in the early 1990s. It was a brilliant concept: All information should be available anywhere to anyone regardless of his or her computing environment. The concept was swiftly implemented at the production and reception ends. For the latter, the browser became the universal access method within a few years of its introduction as Mosaic at the University of Illinois. Since then, innovations in the Internet have come at breakneck speed.

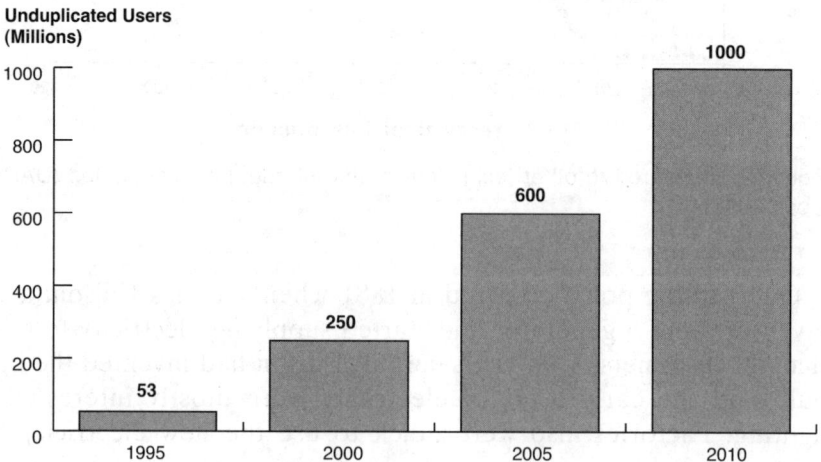

Unduplicated Users (Millions)

Figure 3. Worldwide Internet population forecast 1995–2010 (INPUT 1994)

The fundamental principle of openness has been the basis for the increased rate of adoption of these innovations compared to earlier, proprietary attempts at Electronic Business.

In essence, the Web (with its associated tools such as the browser) provided the application for the Internet that moved it away from being simply a communications vehicle for e-mail traffic. The Web was to the Internet what the spreadsheet and the word processor were to the personal computer: the "killer app."

In 1995, INPUT predicted there would be 200 million users of the Internet by 2000, a number subsequently increased to 250 million. In 1998, the number of regular Internet users was estimated at 147 million, a 240 percent increase over 1996, according to the Computer Industry Almanac. INPUT predicts more than a billion users by 2010.

Threat or Opportunity?

Fear or ignorance often hinders the adoption of new technologies. The Chinese had many of the technologies necessary to have an industrial revolution several centuries before it occurred in Europe. So why did the Industrial Revolution not happen in China? According to Lester Thurow, professor of economics at MIT, China just did not have the right ideologies. New technology was perceived as a threat, not an opportunity, and innovation was prohibited.

In modern China, these types of barriers still exist and IT vendors from Western countries must battle formidable obstacles to enter the Chinese market. But battle they do, for the Chinese Electronic Business market has massive development potential.

With a population exceeding 1 billion, China's Internet users numbered just 1.58 million in 1998, according to the Computer Industry Almanac. That is almost exactly the same number as the 1.57 million people who access the Internet in Finland, which has a population of around 5 million—200 times smaller than China's.

The Internet is breaking down the bonds of time and distance and Chinese officials know they cannot cut the country off from external forces, be they economic, cultural or technological. Despite the high costs of accessing the Internet in China and the restrictions placed by the government, the number of Internet users grew 300 percent during 1998.

Millions

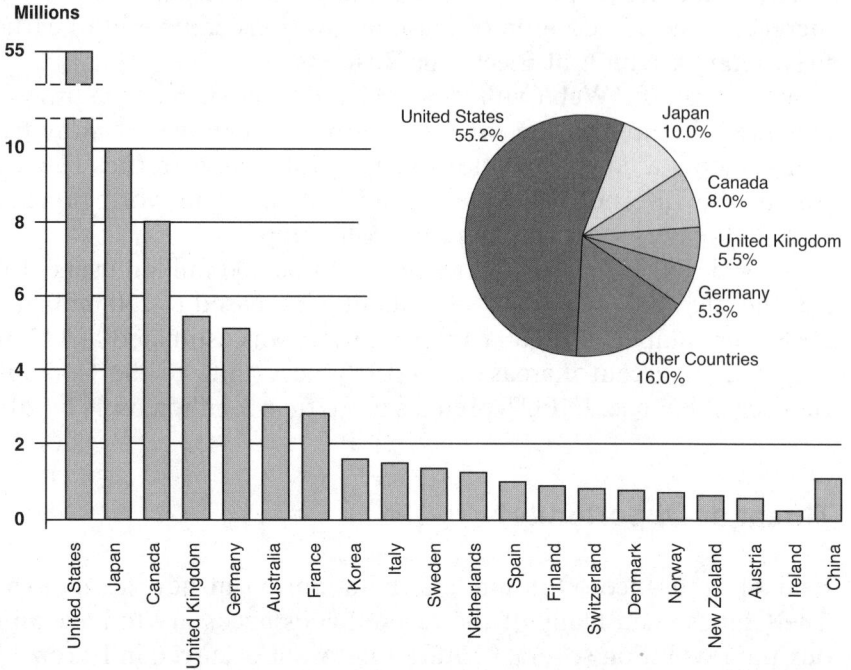

Figure 4. Adults accessing the Internet in selected countries, including China
(OECD 1998)

Once, China had a cultural revolution. Today, it is poised to embrace a digital revolution. Chinese officials know that in order for their nation to continue economic growth at its current rate, they have little choice but to adopt the same policies and tools as the West. And that increasingly means embracing Electronic Business.

The scenario sketched above should send warning signs to those companies in Europe who think Electronic Business is primarily a US phenomenon. They prefer to wait and see and if asked why, they recite a list of reasons: Internet penetration is still low, there is a lack of local-language content, none of their competitors are doing business online, or there are "cultural barriers" that prevent Electronic Business from working in their country.

These are not reasons, but excuses. Once, the Internet could have been written off as a fad that would fade. Today, anyone making such a prediction cannot be taken seriously. The same holds true for Electronic Business. Obstacles exist, but they are all surmountable—as a growing number of businesses and organizations have already discovered.

What is "Electronic Business"?

A simple definition of Electronic Business would be "doing business electronically." That sounds vague but is in fact accurate. The boundaries of Electronic Business are expanding all the time. Electronic Business encompasses the execution of interactive, inter-business processes.

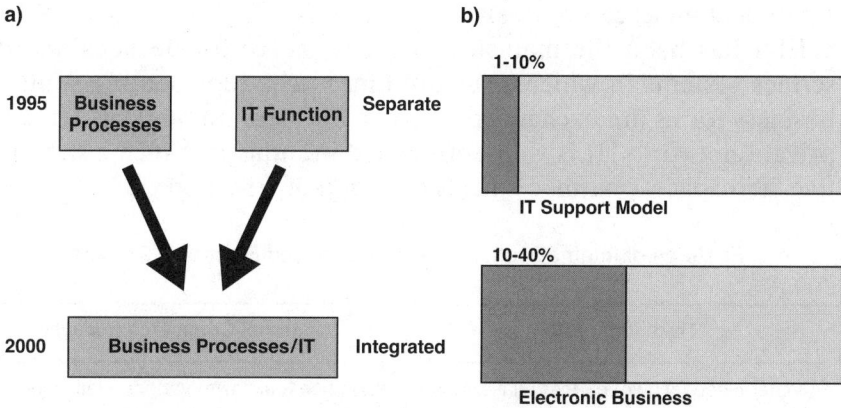

a)

1995   Business   IT Function   Separate
       Processes

2000   Business Processes/IT   Integrated

b)

1-10%

IT Support Model

10-40%

Electronic Business

Figure 5. Integration of IT and business processes (a) and IT costs as percentage of total operating costs (b)

By combining information technology and business processes, Electronic Business re-invents the way we work. It involves the embedding of IT into a business or other organizational process in order to enable that process to operate. Its impact on both internal and external organization gives it the power to create new organizations.

Of course, information technology has long been used to support internal business processes, but Electronic Business is typically more externally focused on the organization's clients and suppliers. IT becomes inextricably part of the process. In this transformation, the portion of an organization's costs that must be spent on IT will have to increase from 2 percent to 10 percent, depending on the industry, to 10 percent to 40 percent in the early 21st century.

Any organization, enterprise, government agency or individual can employ Electronic Business to improve their existing processes or create new ones. In the not-too-distant future, Electronic Business could become the only way of doing business, in some cases.

Electronic Business can be divided into two types depending on whether an organization primarily deals with individuals or other businesses. Despite the media attention devoted to business-to-consumer retailing, it is the business-to-business segment that generates most revenue—and will continue to do so.

Electronic Business is not the same as electronic commerce—although the terms are often used interchangeably. In our treatment, e-commerce is a subset, denoting the buying and selling of goods in a business-to-business setting. It has two components: traditional electronic data interchange (EDI) services and the burgeoning Internet commerce services.

EDI has been the mainstay of e-commerce for decades and describes systems in which structured messages representing standard business forms are exchanged between business trading partners over private networks. It is well-entrenched in industries such as automotive, aerospace, pharmaceuticals and food distribution.

Table 1. Features of Internet Commerce vs. EDI-based Electronic Commerce

Traditional EDI	Internet Commerce (today)
Predetermined (known) partner relationships	Ad hoc (unknown) partner relationships
Closed trading community	Open trading community
Low volume of transactions	High volume of transactions
High-value transactions	Low-value transactions
High security level	Variable security level
Rigid	Flexible
Primary products and services	Secondary products and services

In these sectors, a large company can require its smaller trading partners to use its proprietary system. The advantages in using EDI are primarily for the large company, which can insist that all its suppliers use a standardized invoice, for example.

This traditional type of e-commerce is often dismissed by a new generation of software companies keen to promote business-to-business trading over the Internet. They say EDI was conceived in an era of proprietary technologies and rigid commercial relationships and is thus ill-suited to the complex, fluid relationships that increasingly characterize business-to-business trading. Also, EDI networks have traditionally been costly to establish and use.

Nevertheless, INPUT predicts that the EDI market will continue to grow over the next few years as major companies push its use down their supply chains to ever smaller companies. But traditional proprietary formats for EDI messages will give way to more open formats relying on Internet technologies that will enable all small and medium-sized companies to link into EDI "hub-and-spoke" systems.

The writing is on the wall for traditional EDI. As Internet commerce applications become more sophisticated and the limitations of EDI more apparent, Internet commerce will grow much faster, rapidly eclipsing the older form of business-to-business electronic trading. By 2003, Internet commerce will be four times the size of the EDI market.

These two different worlds of business-to-business trading based either on the Internet or EDI are not mutually exclusive and will in fact merge. In Germany, Siemens Business Services has helped develop a system, called "TECCOM," for the car parts industry. TECCOM uses both EDI and Internet technologies. It helps all parties—workshops, dealers and car manufacturers—by reducing the need to use the phone or fax to make inquiries and order spare parts. Orders are confirmed by e-mail, further eliminating paperwork.

Electronic Business Starts to Flower

For a good demonstration of traditional e-commerce in action, visit the Dutch town of Alsmeer early one morning. Starting at 6:30, buyers from all over the world electronically bid for lots of fresh flowers in the world's largest flower auction. Within three hours, all the flowers are sold. By afternoon, they are already on the way to their destination.

The European plant growing and cut flower industry is big business. Its supply chain is under constant pressure to deliver fresh high-quality products in a time-critical industry. Traditionally, trading involved complicated and expensive logistics, but the use of EDI has recently changed the industry for good.

Until 1997, European growers had to travel hundreds of miles to auctions in order to buy quality stock. Distributed Datanet, an independent organization, was set up to provide an electronic trading system for plant and flower purchasing and distribution businesses.

The aim was to stock garden centers and retailing outlets with minimum cost and time to both trader and buyer.

"We were finding that situations arose where a grower from the US was sending flowers to Europe for an auction, only to have them shipped back by a trader—there simply had to be an easier way," says Hans Klein, managing director of Distributed Datanet. "We decided that the problem was with the physical shipment of products. We needed to implement a more efficient process."

Container Centralen, a non-profit organization owned by Dutch and Danish plant traders, was given the task of simplifying the plant distribution process. Container Centralen proposed a centrally managed electronic trading system that would eliminate the need to move each physical product before it was sold.

Distributed Datanet was set up to support and manage the system. As an independent body, it had no interest in the individual parties in the chain, but simply aimed to enable efficient trading between buyers and growers in return for a small monthly fee.

Container Centralen chose to use EDI rather than the Internet because of security concerns and the need for guaranteed response times. The EDI network is run by GE Information Services, one of the leading EDI players in Europe. Using the GEIS commercial value-added network is more expensive than the Internet, but it offers benefits such as message handling and recording facilities, automatic transmission and reception, and guaranteed message delivery. (All these features are becoming available on the Internet.)

Container Centralen commissioned an independent analysis of the business benefits of EDI. The results were annual savings of around $50,000 on communications with auctions and trades for an average trader.

The Distributed Datanet electronic trading system is a modern-day marketplace. Products are posted onto the system, which is then browsed by buyers. Once an electronic offer is sent, it has to be accepted within the standard one-hour response period. The buyer then makes an acknowledgment, the product is shipped and an invoice sent electronically.

The seamless efficiency of this system cuts costs, speeds up delivery time and therefore improves the quality of the product on arrival at its final destination. Traders no longer need to send employees and products to far-flung auctions.

Without investing in huge IT resources, a large grower can ensure that its stock control system is instantly updated when a trader makes a request. When an order is received using EDI, a message is automatically sent to the stock control system to ensure the right number of plants are in stock.

Once the order is confirmed, the stock control system refreshes its numbers and prevents the plants from being sold twice. The same can be done with shipping and accounting—the systems are linked and the EDI message instigates all the correct documentation to be generated automatically.

Distributed Datanet has been offering the new system to plant growers since the beginning of 1998. It has 25 of the largest traders and 30 growers as participants, meaning the majority of the flower industry in the Netherlands is online. Distributed Datanet hopes to widen eventually to link with another similar system, increasing the business benefits for growing numbers of companies.

Electronic Catalogs: Selling Made Simple

Conducting business-to-business transactions electronically offers several advantages: greater accuracy, faster order processing, lower procurement and operational costs, and better coordination among sales, operations and purchasing. By adopting e-commerce, sellers can become preferred suppliers and purchasers can become preferred business partners. The Internet also becomes an additional channel for sales, marketing and public relations activities.

This last point illustrates the trend in e-commerce to move away from cost-driven procurement to growth objectives. A recent INPUT study shows that the four main goals of adopting e-commerce in the United States now are penetration of new markets, meeting competition, reducing operational costs and reducing order-to-delivery time.

The most important applications in the next years will continue to be procurement-oriented—placing and receiving orders, and making and accepting payments—because the greatest tangible benefits, particularly in large organizations, are seen to lie in this area. The benefits from penetrating new markets or meeting competition are less tangible today.

This whole "e-procurement" area has attracted great attention from the Electronic Business industry. Several vendors offer software that allows buyers or sellers to create "electronic catalogs" in which companies can display the goods they sell—so-called sell-side systems—or details of orders they wish to fulfill—buy-side systems.

Electronic catalogs offer many benefits over traditional sales channels such as printed catalogs and direct sales. Putting a sophisticated sell-side catalog on the Web can provide rich, personalized product information to customers in an efficient, automated process that is available 24 hours per day.

Benefit

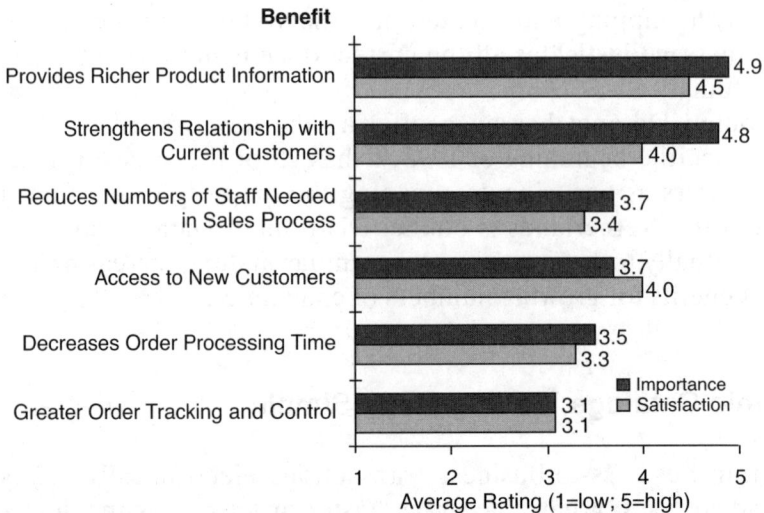

Figure 6. Benefits of electronic catalogs for vendors (INPUT)

Kaut Bullinger, one of Germany's leading office supplies companies, was quick to realize the advantages of this type of Electronic Business. It has been running a highly successful electronic catalog operation covering more than 8000 products since 1997.

"As one of the biggest office suppliers in Germany, we were looking for a way to improve our business processes and develop our business-to-business market using the Web," says the company. Kaut Bullinger turned to Siemens to help it build its online presence and the site was up and running in just three months.

Siemens provided Kaut Bullinger with a hosted solution that meant the company did not have to invest in its own hardware or software. The service links directly into Kaut Bullinger's operational systems, which are based on SAP software.

The use of the online catalog pleases customers and it benefits Kaut Bullinger, which has been able to streamline its "offer-to-order" and "order-to-pay" processes. The savings achieved are passed on to the customer in lower prices.

One of the key achievements has been the optimization of ordering and payment processes for both customer and supplier. Customers no longer need to identify product numbers or their discount terms as the site handles these complexities.

Kaut Bullinger's BTB solution has the advantage that when customers enter their identification code, they are instantly recognized and the prices and conditions shown on the site are dynamically generated to reflect the terms agreed to with each customer.

Such personalized or "one-to-one" marketing is often talked about in the context of online retailing, but as the Kaut Bullinger experience shows, it is just as relevant to business-to-business e-commerce.

Providing richer, personalized service to customers will strengthen traditional business relationships. Unlike printed catalogs, dynamic electronic catalogs display information that is current and complete.

Increasingly sophisticated software can calculate pricing and availability in real-time—or near real-time—for individual business customers based upon pre-arranged contracts.

Also, each business customer can see the products, prices and features unique to his or her company. Add this to the trend in integrating buyer-supplier systems, and e-commerce applications will significantly strengthen existing relationships between trading partners.

The site is operated as a closed business-to-business community and only registered customers are allowed access. The term "extranet" is used frequently to designate such restricted Internet environments beyond the confines of individual enterprises. For our discussion of Electronic Business, however, this book differentiates only between the Internet and Intranets.

Electronic Catalogs: Buying Made Better

On the buy side, electronic catalogs are empowering purchasing departments within companies by allowing better enforcement of business rules, leveraging supplier relationships for price discounts and allowing them to focus on strategic issues.

Benefit

Benefit	Importance	Satisfaction
Reduces Costs of Purchased Products	4.8	4.4
Lowers Procurement/Ordering	4.7	4.4
Richer Product Information	4.3	4.4
Decreases Order Processing Time	4.3	4.1
Decreases Errors/Mistakes	4.0	4.6
Simplifies the Process of Identity	3.7	3.6
Decreases Number of Rogue Purchases	3.7	3.9
Decreases Number of Staff Needed	3.6	4.3
Greater Order Tracking and Control	3.6	3.6
Ability to "Comparison Shop"	2.6	3.0
Access to a Larger Number of Products	2.3	3.3
Access to a Greater Number of Suppliers	1.6	1.5

Average Ratings (1=low; 5=high)

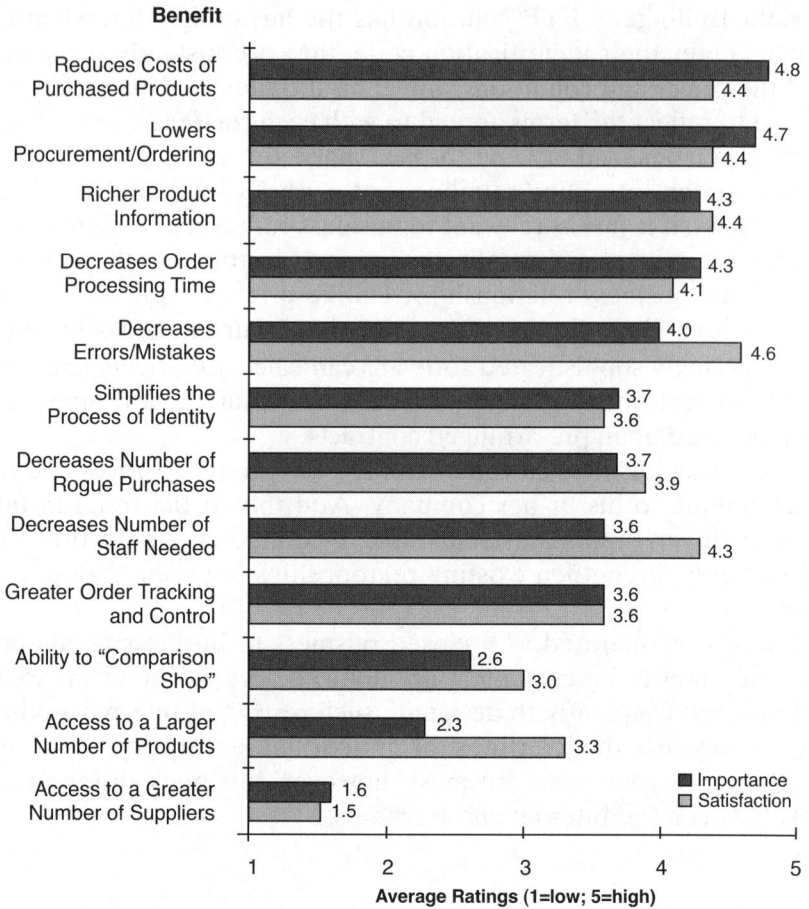

Figure 7. Benefits of electronic catalogs for buyers (INPUT)

The most important benefit by far is reducing the cost of products purchased from suppliers. This happens when a buying organization reduces the number of suppliers with which it deals and promotes strategic partnerships with the remaining core suppliers. As a result, the buyer can leverage the relationship to negotiate significant price reductions on the supplies it purchases.

Suppliers gain from the higher volume of sales and the stronger relationship with their customer. Of course, the buying organization can accomplish this without the use of electronic catalogs, but INPUT research shows that e-commerce applications make it easier to funnel purchases to selected suppliers and this facilitates negotiated price reductions.

An important benefit that is harder to quantify is the effect that buy-side catalogs have on purchasing personnel workload. Since electronic catalogs often allow an individual buyer to automatically check the availability and status of an order, purchasing managers no longer have to spend the majority of their time tracking the location and delivery date of ordered goods. The real benefit gained is from allowing staff to focus on issues such as supplier rationalization and volume price reductions.

One of the primary benefits of using electronic catalogs is their ability to slash long-term costs. Especially within the sales and purchasing functions of companies, electronic catalogs can dramatically reduce costs by improving and automating existing business processes.

Reducing sales costs is achieved primarily by minimizing order entry errors, streamlining the order entry process, and enhancing sales staff efficiency. Like their buy-side counterparts, sell-side electronic catalogs do not necessarily reduce the number of sales staff, but they make the staff more effective given the same amount of resources.

Buy-side applications can lead to significant gains in process efficiencies and reductions in costs. Improving purchase-to-order processes saves costs in several areas. First, improved inventory practices based on shortened procurement cycles reduce inventory carrying costs. Second, the replacement of manual processing of paper forms by more efficient electronic order entry and transaction processing results in lower transaction costs.

Then there are the savings that come from enforcing existing business rules for corporate purchasing, reducing the temptation to purchase from non-approved suppliers—so-called rogue purchasing. Finally, there are the price reductions possible from rationalizing suppliers and thus obtaining greater volume discounts from a smaller supplier base.

Lufthansa, the German airline, plans to run a trial of a business procurement system at its Frankfurt, Germany, headquarters. The project uses software from Commerce One, a US company that has pioneered e-procurement in the United States and hopes to extend it to Europe. The Lufthansa trial will focus on automating purchase of non-critical goods in three areas: office equipment, office furniture and IT products.

The aim is to bypass the purchasing department—a frequent source of bottlenecks—and allow employees to order these goods directly from Web-based catalogs mounted on Lufthansa's Intranet. Orders for standard products below a certain value are sent directly to the supplier, thus bypassing the conventional approval process and saving time, paperwork and the costs associated with manual procurement. The advantage for suppliers is that because the orders are placed online, they are always correct and only for goods that are in stock.

The three areas identified in the Lufthansa trial cover some $10 million worth of orders annually. If the pilot proves a success, online procurement could be extended to a much wider range of goods and services worth $1 billion annually. This is a good example of how Electronic Business changes a process entirely, not just makes it more efficient.

In the short term, many companies hoping to save marketing costs by opening Web commerce sites may be in for a surprise. Instead of saving money, companies often have to initially spend significant amounts of money to advertise and educate customers about new e-commerce applications. Increased awareness will eventually drive customers to Web sites and electronic catalogs, but there will be a lag between initial expenses and eventual sales.

Falling Barriers, Falling Prices

The Electronic Business industry paints an enticing image of Web-based procurement in which the combination of electronic catalogs and Internet search engines will create a virtual market similar to today's stock exchanges in which procurement operates according to a dynamic pricing model whose efficiency ruthlessly drives down prices.

This will lead to the commoditization of many products in which only those businesses offering the lowest prices will survive. The theory assumes the Internet can provide ubiquitous access to product and marketing information, and so eliminate many of the traditional barriers that ensure that today's physical markets are far from perfect.

This theory also assumes that products are interchangeable and customers will buy from any vendor. For industrial goods and services, this is often not true. One of the most important recent trends in industries such as manufacturing is the growth of "preferred" suppliers that benefit from long-term, privileged relationships. They may, for example, be given access to the customer's demand planning systems to allow them to optimize their own production.

The desire for closer, more direct contact with customers is one of the most common reasons for adopting Electronic Business.

Boeing is using the Internet to give its airline customers access to technical information needed to maintain the world's largest fleet of jetliners. Using a password-protected, secure Web site, airlines can order and track spare parts shipments. (Note that security requirements for aircraft parts are far greater than for flowers—yet Boeing uses the Internet, while Centralen referred to earlier does not because of "security"!)

By the end of 1998, most operators of the more than 10,000 Boeing airplanes in service worldwide were using the Internet or dedicated network facilities to access key Boeing support services.

In 1998, the site handled 1.6 million transactions—more than double the volume of the previous year. Boeing launched the PART Page in late 1996 as a means for smaller airlines in particular to benefit from Electronic Business without having to invest in costly mainframe systems.

It now accounts for well over half the volume of all transactions received by the Boeing spares organization, dramatically reducing the traditional reliance on phone, fax, telex or mail.

Boeing also allows its customers to retrieve technical drawings, service bulletins, maintenance manuals and other vital maintenance data as electronic files. This service, called Boeing On-Line Data (BOLD) is claimed to be faster and more accurate than conventional paper or microfilm-based retrieval methods.

BOLD users have real-time access to Boeing databases using standard computer workstations linked via high-speed, wide-area-network providers. Last year, 65 airlines—including the world's largest carriers—generated 9 million BOLD transactions, twice the volume of the previous year. The number of customers using BOLD grew by more than 50 percent in 1998.

"More and more of our customers are taking advantage of our digital online systems," says Rich Higgins, vice president of maintenance engineering at Boeing's commercial airplanes group. "Ultimately, we want our customers to be able to get all of the information they need to operate their Boeing aircraft through a single network connection."

Banking on the Internet

The retail financial services industry deserves a special place in any discussion of Electronic Business. Banks are among the most intense users of technology and they are involved in some of the more exciting developments on the Internet such as secure payment technologies, transaction-enabled Web sites and advanced customer relationship management.

Many banks have identified Internet banking as a key route to increase their market share and retain customers. More than 1000 US banks have or plan to have fully transactional Web sites.

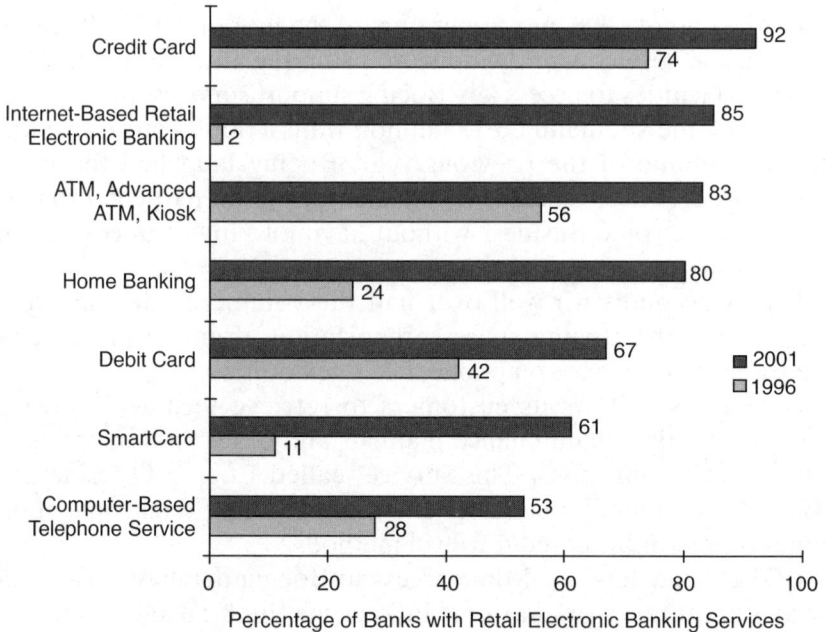

Figure 8. Electronic banking services in 1996 and 2001 (INPUT)

European banks have also been quick to embrace Electronic Business as a competitive weapon. The impact of the euro and lower margins has unleashed a wave of consolidation in Europe's retail banking industry in which Electronic Business is playing an increasingly important role. Banks recognize that customers are looking for easier ways to access information and conduct transactions; they see the Internet as a major commercial opportunity. The most important financial reason for this interest in Electronic Banking is added revenues, but saving money is a close second. Branch-based transactions are costly and the extensive branch networks that allowed banks to expand in the past are difficult to sustain in an era of intense competition. Call centers and automated teller machines have helped drive down the cost of basic transactions, but the Internet has the ability to radically reform their cost base.

Already some European banks are offering cheaper mortgages or better savings rates to those customers who forsake traditional branch-based banking and agree to only use their account via direct channels such as phone banking or the Internet.

A good example of the new generation of virtual banks is Germany's NetBank. Siemens Business Systems designed and built the Internet site for NetBank, which belongs to the Sparda Bank group.

It aims to offer services that go beyond the usual financial offerings and allow its customers to get the most out of the possibilities of Internet banking. As well as offering account information around the clock, NetBank offers personalized information tailored to each user, up-to-the-minute news, and electronic shopping for products ranging from videos and CDs to pension plans.

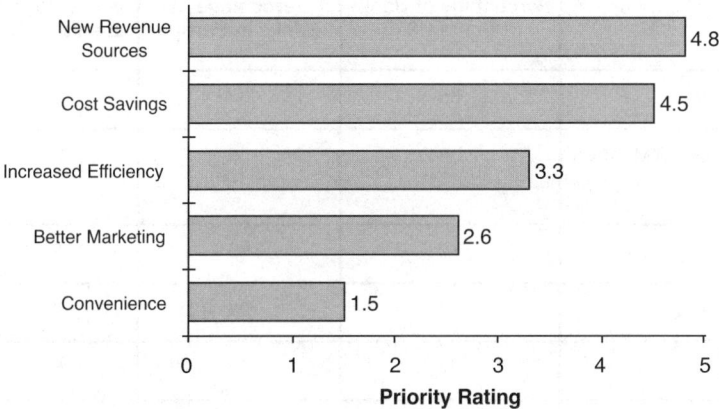

Figure 9. Electronic banking benefits (INPUT)

Another SBS customer, Bank Austria, has implemented a similar service. It has been voted Austria's best Internet bank in an independent survey carried out by the Lafferty Group, a UK financial research firm.

Bank Austria is the largest financial organization in Austria and one of Europe's top 40 banks. Despite this leading position, the bank knows that its future competitiveness depends on how well it can accommodate changing customer demands with innovative products and services.

"The integration of the bank's sales delivery channels is a key element of that strategy," says Robert Macho, assistant channel manager for online banking at Bank Austria.

Each online customer gets a personalized home page generated dynamically when the customer logs on. This replaces the bank's previous static pages and is seen as an important tool to create a stronger personalized relationship with each online customer.

Best of Both Banking Worlds

But most Europeans are not yet on the Internet and so banks must be careful to not alienate the majority of customers that still want to walk into a branch. This is the conundrum facing Deutsche Bank, Germany's biggest bank, which also operates a successful direct channel, Bank 24.

Table 2. Use of e-banking services in the United States, Europe and Asia in 2001

Retail Service or Product	Percentage of Banks in the US	Percentage of Banks in Europe	Percentage of Banks in Asia
Home Banking	85	82	80
ATM, Advanced ATM, Kiosk	88	86	82
SmartCard	70	69	54
Credit Card	99	93	86
Debit Card	72	74	60
Computer-Based Telephone Services	64	58	36
Internet-Based Remote Services	50	41	23
Internet Use For Retail Electronic Banking Services	90	85	72

Deutsche Bank has more than 6 million customers while Bank 24 has around 400,000. The bank has proposed merging the two operations to offer the best of both worlds: a wide range of products, services and personal advice available through its extensive network of 1450 branches combined with the constant accessibility and special services that Bank 24 offers through its telephone and online services.

Nearly 6 percent of all private banking customers in Germany had an account with a direct bank at the end of 1998. The bank recognizes that for an increasing number of people, direct delivery channels are an important criterion when choosing their bank.

At the same time, the bank's research shows that, in the medium term, most customers want the best of both worlds: the same right to use the advisory and familiar services of the branches and the convenience direct banking offers. Deutsche Bank Group is ideally placed to deliver this combined offering.

Pioneering banks in regions with a less-developed banking industry are also adopting Internet banking. One example is Turkey's Garanti Bank, which claims to have been the first Turkish bank to offer online services to its entire branch network—no mean feat in a country twice the size of Germany.

Garanti Bank sees the Internet as a strategic delivery channel and hopes to reach 100,000 users within the first three years. The bank outsourced the technical implementation of its electronic banking site to Siemens Business Services, which created a new Internet service provider (www.garanti.net.tr) to provide high-speed access for the bank's customers.

Insurance

In a recent survey, a group of insurance executives, when asked about their Electronic Business objectives, rated improving customer services their top priority. First, they wanted to inform customers about the insurance products they had purchased. The next two most important areas were the executives' ability to more rapidly introduce new services and to reduce expenses.

Initially, the Internet in insurance will serve as an accelerator in the agent/broker relationship between the customer and the insurance organization itself—especially in the area of obtaining new business. Databases that enable agents to quote information online and transmit policies from various offices are being developed. In this way, the time taken to implement a policy has been reduced from weeks to days and, in some cases, hours. Generally on the payment side, there is acceleration in the claims processing area. Although this is to the benefit of customers, it can significantly accelerate outflows from insurance companies.

E-mail is also being used for many insurance activities that are not transaction-based. It is being used to communicate with customers,

agents at other offices or agents working at home. Insurance companies are also marketing their services more aggressively via the Internet. They are providing facilities for customers to get quotes and
access to other services online. Initially, these services are addressed
to their existing customers, but they will rapidly spread to prospects
as well. Established companies are being forced in this direction even
though this threatens to introduce competition with their prime sales
channel of agents and/or brokers. Competition is the forcing factor
here where direct marketing organizations, several of which started
in Europe, are penetrating targeted segments of the insurance industry such as automobile insurance.

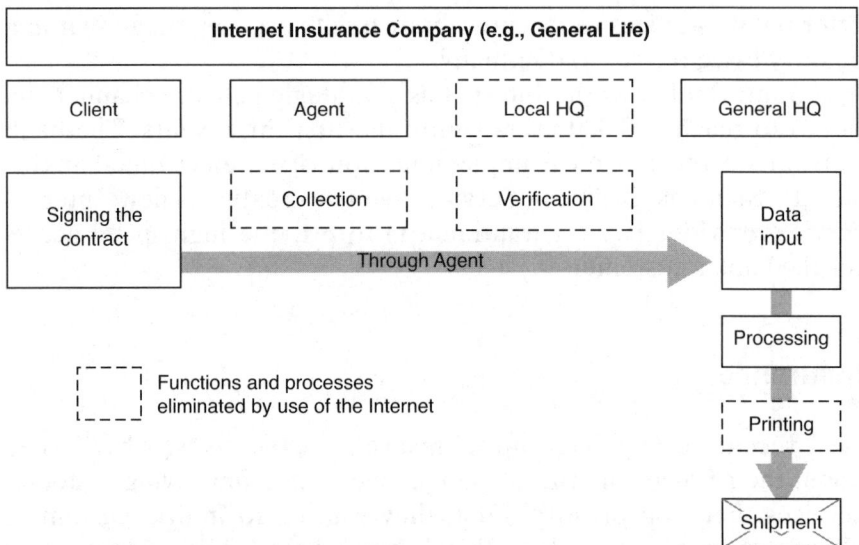

Figure 10. Online services can reduce the time needed to process and ship an
insurance contract from weeks and days to hours.

Technology is supporting these trends. The rapidly changing type
of devices used by insurance salespeople is a prime component of this
technology. This, for example, has led to the deployment of several
hundred thousand handhelds on a worldwide basis, with probably
about 100,000 in the United States. The first step involved agents' use
of non-communicating laptops. They encountered problems with
laptops, however, particularly in high crime areas, and also with limited battery life at a client's location. Some companies have switched
to palmtop-sized devices such as the Casio with a touch screen, Windows and printers. Such devices, particularly for agents operating in
low-income areas, must be rugged, connected through wireless, sim-

ple and cheap. They also require a minimum eight-hour battery life. Examples of these devices are the Psion 3A being used by Prudential in the United Kingdom and Casio by Pearl Insurance. People's Security has moved to a Verimation "memo" system and PMS to VI link.

Another good example is CRS Underwriting, which is using a handheld with a pen stylus, interface, digital camera, modem and keyboard interfaced with a device using AA size batteries. Such systems provide fast access to file details of customers and available policies, support for claims processing and insurance adjusting.

The insurance industry has been slower to embrace Electronic Business in Europe, held back partly by the complex regulations for selling insurance. But this situation is changing, particularly with regard to simple insurance products such as motor, home or travel insurance. One of the European leaders in online insurance is Eagle Star of the United Kingdom.

The Eagle Star Direct site was launched in August 1997. In its first year, it attracted more than 500,000 visitors, with more than 100,000 going through the full process to obtain quotes for motor insurance. The company sought to make the online form-filling process quick and simple and ensure that quotes could be delivered within eight seconds of their hitting the quote button.

The benefits of moving to virtual distribution channels seem irrefutable. Products such as life insurance and mutual funds are typically sold through a network of agents and intermediaries whose salaries can make up 70 percent of total operating costs. These costs have been spiraling out of control for many years. In the United States, for example, the return on investment for an independent agent was just 3 percent in 1996 compared with 13 percent in 1980, according to Computer Sciences, the IT services company.

Investing Online

For Europe's retail investment industry, the challenges of selling online in Europe are perhaps even greater than for insurance companies. In much of continental Europe, investment products are sold primarily through retail banking channels and depend heavily on face-to-face contact. Nevertheless, there are signs that even this traditional distribution channel will not remain immune from Electronic Business.

Despite the fact that many financial services companies have cut their cost base and tried to raise productivity, extensive physical distribution channels have become an expensive liability in an age of virtual finance.

The rapid growth of online trading has shown the financial services industry a radically different way of doing business, one in which the client does much of the work and in return pays lower charges.

In the United States, there are clear signs that online investing is a mass-market phenomenon thanks to the success of companies such as Charles Schwab and E*Trade, which have pioneered low-cost share trading on the Internet.

E*Trade started life in 1996 and now has more than 500,000 members who receive free access to investment tools and research. The company's trading volume at the end of 1998 averaged about 43,000 trades a day. The success of E*Trade has encouraged other online brokers and together they accounted for around half of the retail share trading market in the United States in 1998.

The extraordinary growth of this market has forced traditional full-service brokerages, such as Merrill Lynch, to also embrace the Internet, although they are keen to avoid competing in a price war with the discount Internet brokerages.

Europe also has a growing number of online brokerages. They include US pioneers, such as E*Trade and Charles Schwab, who seek to expand into new markets as well as home-grown players.

In 1997, Charles Stanley, a London stockbroker, launched one of the first online share trading services in the United Kingdom. The Xest service enables customers to invest in the UK stock market for a flat fee of £20 a transaction. Xest accepts funds to settle transactions in a range of currencies, including all European denominations, and is allowed to accept orders from most parts of the world, including all member states of the European Union. The US online brokerages, by contrast, are limited to US residents.

Clients benefit from fast and efficient electronic settlement while continuing to receive shareholder rights directly from the company. Xest achieves this by using the London Stock Exchange's paperless settlement system, called Crest, which overcomes the need for investors to physically hold certificates. In addition, investors receive the latest company and financial news and real-time prices on all shares listed on the exchange.

Ameritrade, a US brokerage firm, has struck a series of alliances with several European brokers to create online trading sites. One of its partners is Bank24, the direct banking arm of Deutsche Bank, which has announced plans to make new equity issues or initial public offerings (IPOs) available to investors via the Internet.

Traditional	Banking	Brokerage	Lending	Insurance
Bricks and Mortar	CitiGroup Wells Fargo	CitiGroup Merrill Lynch	CitiGroup Countrywide	CitiGroup Allstate
Online Extension	e-Citi e-Wells	e-Citi Prudential Online	e-Citi Countrywide	e-Citi Allstate

Vertical ECBs

Traditonal Providers		Discover (MSDW) Schwab	First USA	
New Entrants	Telebanc Net.B@nk	E*TRADE Ameritrade		IHA

Destination/Portal

Financial Destination	◀———— Destination E*TRADE ————▶
	Intuit

Horizontal Portals	◀———— Yahoo! ————▶
	AOL

ECB   E-Commerce Bank

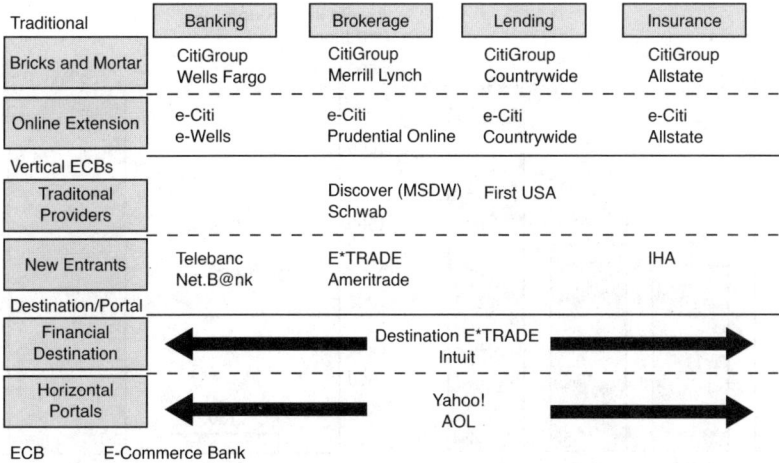

Figure 11. Online financial services landscape in the US (Charles Schwab)

Deutsche Bank claims to be the first big European bank to offer this facility, which reflects the growing interest in share ownership on the continent, and in Germany, in particular. Most of the companies that Deutsche Bank will take public will be joining the Neuer Markt, the new German exchange for high-growth companies modeled after the NASDAQ in the United States.

All Internet users can access Deutsche Bank's comprehensive information offering with lists of past and forthcoming new issues, sales prospectuses and a company profile. But the online subscription facility is limited to customers of Deutsche Bank who have online account capability.

The bank expects the proportion of private investors who send their IPO orders to the bank on the Internet to increase rapidly. For issuers, the Internet has the advantage that they can provide company information more easily than before, not only to the institutional investor but also to private investors.

Fidelity, a leading US retail investment house, is one of the trailblazers in this field. Its US Web site allows US investors to open an account online and trade not just mutual funds but also shares, thus competing with the online brokers.

So far, most trades placed by online investors have been stock trades. But the value of mutual fund assets managed on the Internet could exceed $1 trillion in the next five years, according to industry estimates.

Customers

**Assets in Accounts
($ Billion)**

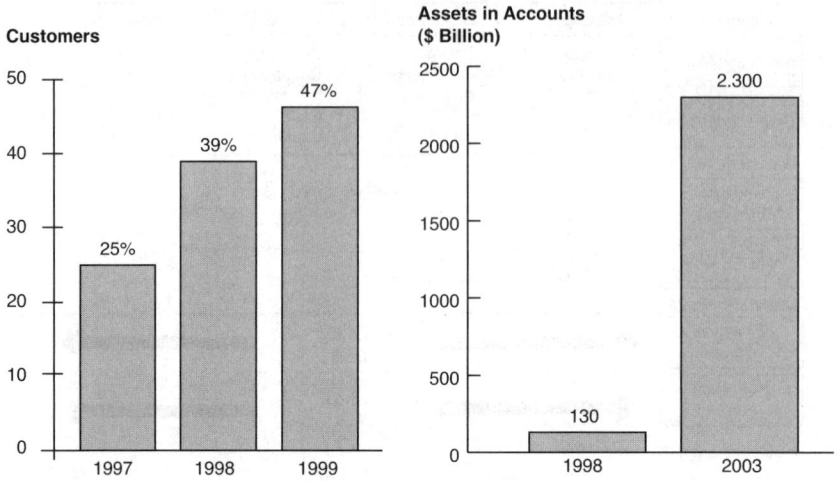

Figure 12. Percentage of Charles Schwab customers using e-trading (Charles Schwab) and assets in online brokerage accounts in the United States, 1998–2003 (INPUT)

Fidelity has also created Web sites for online investors in Germany and the United Kingdom. These are less sophisticated than the US site and limited to existing customers who want to buy and sell Fidelity funds online.

Electronic Business presents particular challenges for Europe's retail investment industry because of the different regulatory regimes, tax treatments and product ranges of each country. Despite the barriers and the low levels of Internet penetration in many European countries compared with the United States, there is growing interest in developing and expanding the online financial services industry in Europe.

Online Retailing

For many consumers, the principal attraction of the Internet lies in its promise of cheaper shopping. Books, CDs and travel have been the areas that have benefited most from the greater price transparency that the Internet brings. The same thing is starting to happen with other consumer products, even big-ticket items such as motor cars.

One need only consider the significantly different prices charged for identical car models across Europe to appreciate the possible benefits that would come from the creation of a single "virtual" market in Europe's car industry.

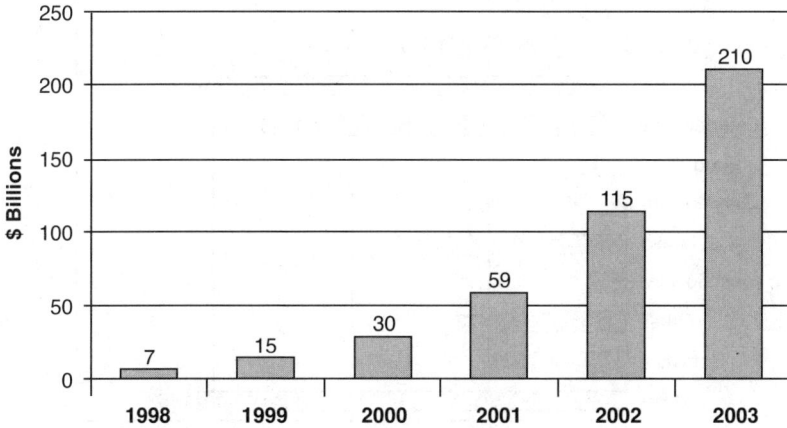

Figure 13. Worldwide value of e-retail (INPUT)

With the advent of the euro, Europe's car-makers can no longer hide behind the excuse of currency exchange rates. The European Commission conducted an investigation into car prices and discovered that most of the difference between final "sticker" prices in different EU countries could not be justified.

Comparison shopping for products such as cars can be difficult and time-consuming. A buyer first must identify and locate the local dealers of a particular brand; call them to see if they have the desired model, and if not, how long it would take to arrive; and then negotiate a price with each dealer.

US car buyers are turning to the Internet to find the best price for their desired model. Just a few hours spent comparing cars and prices on the Internet can produce big savings. This also allows them to delay the moment when they have to confront a car salesperson—for many, an unpleasant experience.

One US company, Autobytel.com, has created a highly popular Web site for car buyers. The company is expanding internationally. It has launched Autobytel.se in Sweden and was preparing to launch a UK version, called Autobytel.co.uk, in mid-1999. It also plans a Japanese version of the site.

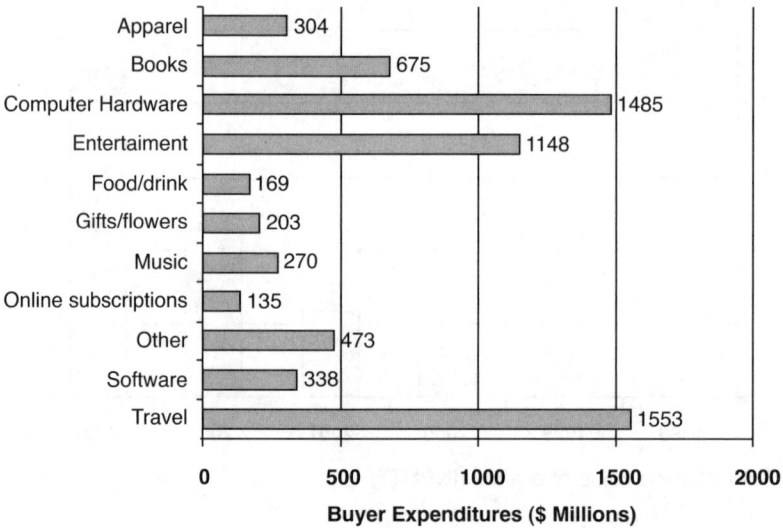

Figure 14. Worldwide value of e-retail by product category in 1998 (INPUT)

Many potential buyers are attracted to Autobytel because it eliminates much of the hassle and uncertainty in buying a car. Customers enter the make and model of car they want, how much they are willing to pay, and whether they would like to buy from a dealer or a private seller.

They then tell the site how far they are prepared to travel to buy the car. Within seconds, the potential buyer can view an exhaustive list of all the cars for sale within a given distance from home, including prices, photographs and detailed descriptions of each vehicle.

After deciding which car to buy, the customer enters the ZIP code where he or she lives and the make and model of the car desired. A screen pops up requesting further details such as the desired exterior and interior color schemes or the size of engine.

Then the customer completes a new car purchase request, selecting which manufacturer options such as radio or anti-lock brakes, to include on the car. With these selections and some contact information for the customer, the request goes to the Autobytel dealer closest to the customer's home. Within 24 hours, the dealer contacts the customer with a fixed price, thus eliminating the traditional and—for many buyers—unpleasant process of haggling over price. Even financing and insurance can be arranged online.

The dealer explains the options and all the paperwork is prepared before the customer arrives to complete the transaction and pick up the car. The customer is also asked whether or not he or she wishes to acquire service agreements or after-market products available from their dealer.

Autobytel.com launched its Web site in 1995. In its first nine months of operation, the company processed 43,000 purchase requests. In the first three months of 1999, that number leapt to 489,000 and the service was handling more than $24 million in sales each day. The company claims to have helped more than 2.5 million car buyers since its inception.

Autobytel makes it money by charging dealers a sign-up fee and a flat monthly fee, ranging from $995 to $2500, regardless of how many customers it sends to the dealer. Strictly speaking, it is an information service, not an Electronic Business.

The attractions of buying a car this way have spawned the creation of similar services such as CarsDirect.com, an online venture backed by Michael Dell, whose company, Dell Computer, pioneered selling PCs over the Internet.

The mainstream car industry is waking up to the challenge that Internet sales pose to its traditional business model. Such Web sites give buyers control over the car-buying process, shifting the balance of power in the industry. If the "virtual showroom" takes hold, do carmakers need extensive—and expensive—networks of physical showrooms and dealers? Companies such as General Motors are actively reconsidering what their role and the roles of their dealers should be in the Electronic Business world.

Once the buyer has chosen his or her car on the Internet and closed the deal online, the dealer's role is reduced to delivering the car to the customer. Might it not make more sense to have the buyer deal directly with the car manufacturer, who can then ship the car straight from the factory to buyer? In this model, the dealers evolve into service organizations. The first steps in this change are becoming evident in the United States with the purchase and consolidation of dealerships.

The Disappearing Middlemen

This ability to save costs and time by eliminating the intermediaries in the distribution chain goes by the name of "disintermediation." In the early days of Electronic Business, disintermediation was seen as a powerful threat to many businesses. Proponents cited the example of Dell, which revolutionized the PC business with its famous build-to-order model and a direct sales operation that bypasses traditional PC dealers.

Other PC makers have attempted to copy Dell's model and have built Web sites to take retail orders for PCs that are then shipped direct to the customer. But Dell's speed and efficiency in turning an order into a delivered product remains unmatched.

To fans of disintermediation, a car is much the same as a PC, so why not eliminate the dealers and order cars direct from the manufacturer? Assuming the customer knows what brand and model he or she wants, the buying process can be reduced to checking the boxes for the desired options, entering payment details and specifying a delivery address.

But cars are not the same as PCs. For one thing, cars need regular servicing (although this is becoming a less frequent requirement), and so customers may still need to regularly enter a dealership even if it is only to drop off their car for servicing and pick it up later.

In an increasingly virtual world, this physical contact is a marketing opportunity the car industry is keen to maintain. Volkswagen, for example, has given its showrooms a more customer-friendly, service-oriented image using multimedia kiosks from Siemens.

The PC-based system, called Service Auto-Mat (SAM), automates all elements of vehicle servicing from handing the vehicle over right through to pickup and payment. The benefits for the customers are a 24-hour hand-over and pickup service with faster processing of servicing formalities. This in turn allows the Volkswagen dealer to handle more cars at peak periods.

Thanks to its built-in multimedia capabilities, the SAM kiosk can not only process servicing formalities, but present the customer with a wide range of marketing information and other details: information on VW accessories, on the VW Club, leasing and finance offers, insurance services, dealer-specific news and offers on used cars.

This is a good example of the potential of kiosks and smart ATMs. However, the kiosk material must be coordinated with what is on the Web, and the Internet must be used to refresh the material in each kiosk. Other methods are too slow and ultimately too expensive, except for video and graphic information, which is still too large to be comfortably distributed via the Internet.

The Internet is only starting to impact the business of selling cars. But one industry in which the Internet has already had a profound effect is book selling, even though the volume of books sold this way is as yet very small.

The Amazon Skyrocket

Who has not heard of Amazon.com, the US online bookstore that has become the archetypal Internet commerce success story? In just four years, it has grown from nothing into a major retailer with 8 million customers in more than 160 countries. Its sales grew more than 300 percent in 1998 to $610 million, and while it has yet to make a profit, its share price has rocketed.

In mid-1999, Amazon.com was capitalized at more than $30 billion, making it the fourth-largest retailer in the United States by capitalization. The secret of Amazon's success? A simple formula based on discounts of up to 40 percent from regular bookstore prices, efficient service and a now-famous brand name. Jeff Bezos, the company's 35-year-old founder, deliberately chose a simple name that began with 'A' to ensure that the store's name could be remembered.

"If you want to go to McDonald's and have a hamburger, you never have to know how to spell 'McDonald's' to get there," he once said. "But to get to Amazon.com, you do have to know how to type the name correctly into your browser."

Today, Amazon's Web site gives access to a catalog of more than 4.7 million books, CDs and computer games. But for the company to keep growing, it knows it must expand into other areas. In early 1999, the Seattle-based company bought stakes in online companies such as Drugstore.com and moved into other businesses, including auctions and electronic greeting cards.

The aim is to draw more customers to Amazon's Web site and so become a destination site where customers can buy a wide range of products apart from books. This strategy is quite a departure for Amazon and it puts it on a collision course with several other companies.

One such company is eBay, the Silicon Valley company that invented Internet auctions and whose market value now surpasses that of Amazon. In early 1999, the gross value of the goods sold through the eBay Web site was running at more than $2 billion a year—double that of Amazon. Unlike Amazon, however, eBay acts as a broker rather than a retailer and its revenue comes from the commission it charges on the transactions that pass through its site.

In gross revenue terms, eBay is dwarfed by Amazon, but in some respects the online auction company has a more interesting and radical Electronic Business model. Similar to the Distributed Datanet flower auction mentioned earlier in this chapter, eBay's Web site functions as a truly electronic marketplace, an application that would be impractical—if not impossible—without the Internet.

Since eBay started, several online auctions have sprouted in Europe. The most famous is probably QXL, which was set up in the United Kingdom in 1997 as Europe's first online auction house. The company's revenues are currently small—the run-rate for 1999 is around £10 million—but it is growing in excess of 20 percent a month. In October 1998, a German version of the site (www.qxl.de) was launched, and a localized French site (france.qxl.com) followed in December 1998.

Since its formation, more than 50,000 auctions have been hosted on the QXL site. It runs private person-to-person auctions as well as auctioning job lots of branded goods such as computers, electrical and household goods, and, most recently, airline seats. In February 1999, QXL received a $12 million investment from four leading venture capital funds in what was claimed to be the largest investment in a European Internet company to date.

The workings of an online auction house are simple. Users advertise the goods—mostly secondhand—they want to sell and other users enter bids. If a bid is knocked out, the bidder is notified by e-mail and given the chance to up his or her bid until the auction has run its course and a winner declared.

The auction site makes its money by charging commission on each transaction—6 percent in the case of eBay. This is a very different model from online retailing and it has advantages. For example, unlike Amazon, eBay's customers do all the work—clearing payment from buyers, arranging delivery, etc. eBay is also profitable.

Amazon's chronic losses are sustainable because investors continue to believe Bezos' message that continuing heavy investment is necessary upfront to establish a powerful retailing brand on the Internet and ensure future profitability.

Much of Amazon's investment has gone into "behind-the-scenes" areas such as expanding its physical distribution network to accommodate growth in customer orders. Developing an efficient system to accept and fulfill orders and returns has become critical to Amazon's business.

Amazon is also expanding in Europe. In 1998, it opened online stores in the United Kingdom and Germany. In early 1999, it announced that it would open a site for Spanish-language books in Spain.

The UK site (www.amazon.co.uk) is based in Slough and carries a catalog of more than 1.2 million titles from UK publishers, along with fast and easy access to more than 200,000 US titles.

The German online store (www.amazon.de) is headquartered and has a distribution center in Regensburg, with editorial and marketing offices in Munich. It was launched with more than 400,000 titles from German publishers, as well as fast access to nearly 500,000 US titles. These aggressive expansion plans must serve as a wake-up call to European booksellers.

Bezos was not the first to start a major online bookstore in the United Kingdom. That honor falls to Daryl Mattocks, the 34-year-old founder of bookshop.co.uk, which trades as the Internet Bookshop. He started planning the Internet Bookshop in 1993, working from a laptop computer, and the system went live on the Internet in 1994.

Four years later Mattocks sold it to WH Smith, a large UK chain of book stores, for almost £10 million. Today, the Internet Bookshop has more than 1.4 million books in its database and aims to supply all UK-published books at the cheapest price available on the Internet, matching competitors prices if necessary.

One of the problems that the Internet Bookshop has had to face is that the prices for US editions of books are often cheaper than the UK edition produced by a British publisher. Net surfers soon learned to check the price of a desired book on Amazon.com before buying from the Internet Bookshop. Even with the additional cost of trans-atlantic shipping, often it was cheaper to buy the US edition from Amazon in the United States than the UK edition from the Internet Bookshop. This is an excellent example of Electronic Business breaking down geographic barriers. We will further discuss this point later.

Books, like many goods sold on the Internet, are essentially commodities. Most customers will surf across to a different Web site if they think they can get their book at a cheaper price. Indeed, "comparison shopping" is one of the big threats to big-name online retailers because it threatens to undermine customer loyalty. Amazon has unusually high customer loyalty—some 66 percent of its customers are previously registered users—and its efficiency in fulfilling orders is legendary.

Nevertheless, price competition in all areas is increasing, particularly with the growth of comparative shopping sites that boast they can find the best deal for shoppers and thus save them the task of visiting each individual online store.

One such site is Shopper.com. It offers a free computer product and pricing search engine that compares more than 1 million prices on more than 100,000 products from 100 cyberstores.

Traveling On

The travel industry is a very large sector. Out of the world gross domestic product (GDP) of approximately $27 trillion, the travel industry accounts for at least $2 trillion. With a cumulative annual growth rate of 7 percent, it is expected to double over the next 10 years. AMEX Travel Services estimate that the travel industry will reach $2.5 trillion by 2002, out of a global GDP estimated to be then worth $31 trillion. US firms alone spent $156 billion on travel in 1996, according to AMEX Travel Services. So travel is poised to become the biggest business-to-consumer market on the Internet and one of the biggest BTB markets.

Most online travel companies focus on the leisure market and must attract large numbers of customers because of the low margins on selling discounted tickets. They must advertise on heavily-visited "portal" sites such as Yahoo or Lycos. But such sites are expensive for advertisers. For BTB companies this approach is not cost-effective.

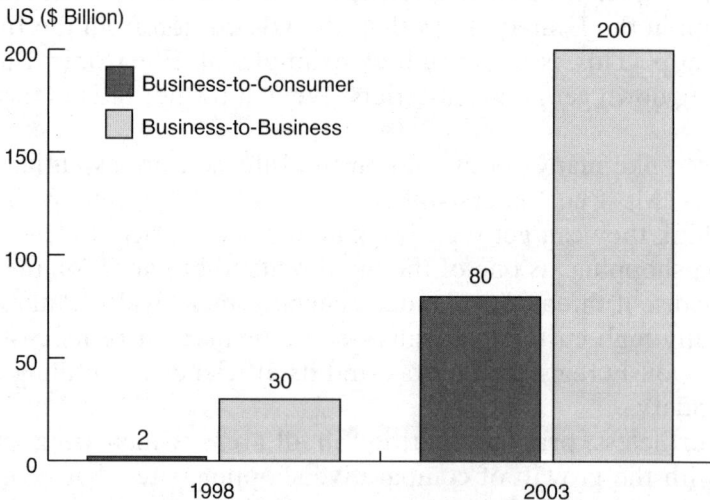

Figure 15. Worldwide value of travel purchased electronically 1998–2003 (INPUT)

Instead, they focus on the small-business market and have copied the approach of leading online US business travel companies, which concentrate their marketing on business travelers, offering them a range of cheaper products and specialized functions on their Web sites. One advantage they have is that they can advertise on sites that are more specialized and less expensive than the mainstream portals.

Moreover, by focusing on the business market they can sell advertising at premium rates on their own sites because of the attractive demographic profile of business travelers. Experience in the United States has shown that significant revenue can be earned from advertising and the sale of leisure products to the existing users of the sites.

These new companies target small and medium-sized businesses because large European companies typically already have exclusive arrangements with specialist business travel agencies. While some of these agencies also have online sites, they are only accessible to their customers.

Everything Works Online, Except for the Mall Experience

Travel and books were among the first products to be sold online. Yet almost all consumer products that do not require a complicated buying process or physical contact with the product can be sold online. Fashion is problematic, as is food, but even these areas contain opportunities.

Cedlerts Fisk, a small Swedish luxury food supplier, offers a good example of how the Internet can be put to imaginative use. Traditionally selling just to restaurants, Managing Director Christer Öholm hit upon the idea of using the Internet to also sell lobster and salmon to consumers.

"I was convinced that it could provide us with a low-cost and highly practical means of reaching new markets," he says. Selling lobster over the Internet has its problems: Cedlerts Fisk had to design special packaging and ensure delivery within 24 hours.

But Öholm says the effort is worthwhile because the Internet has become a powerful marketing and sales tool for the company that can, for example, learn customers' preferences and track buying patterns.

On the Internet, the consumer—not the merchant—is in control. Sellers know they must make the experience worthwhile and enjoyable if customers are to revisit their site.

A report on e-commerce from eMarketer, a US market research firm, identifies the top five factors that motivate consumers to buy online: convenience, security, customer service, variety and price.

Customer expectations about the speed, helpfulness and comprehensiveness of customer service are being raised daily, and while the European online retailing market is less developed than in the United States, it is only a matter of time before European online consumers start to become as demanding as their US counterparts.

One area of e-retailing that has been tried and failed is the "virtual mall." The virtual mall was a concept introduced in the early 1990s. It suggested that an electronic retailer—a telecommunications company, for example—could set up a Web space, a virtual mall, that would collect a dissimilar set of vendors—banks, manufacturers, publishers, and the like—who wanted to put their storefronts into the mall.

The mall owner would charge a transaction fee for bringing traffic to the store. Even Microsoft tried and failed with this one. The reasons for its failure are symptomatic of the difficulty of transferring physical world reasoning to the electronic world.

Virtual mall owners forgot to ask why people visited shopping malls. The answer, fundamentally, was convenience: ample parking and easy access to a variety of shops so that the shopper could buy a variety of goods without having to drive or walk far among stores.

But in the electronic world, such convenience has no meaning. If we want to look at cars, we can look at all varieties with a few clicks of the mouse. There is no value in having easy access to a few dissimilar organizations. In fact, it is counterproductive. Also, store owners quickly realized they could link their mall presence to their non-mall presence and avoid having to pay a transaction fee to the mall owner.

What is very useful and what INPUT predicted would be successful is the "information mall," now commonly known as the "portal." Portals bring consumers and companies to their sites to provide them with a rich and varied set of information about a wide range of subjects or one specific subject, depending on the portal characteristics.

So if you are looking for information on mortgages, for example, you could sign on to a portal site that will enable you to look at all kinds of mortgage offerings, provide information on vendors, rules or basic rates and enable you to shop around for the best deal.

There are even portal sites for general practitioners in the United States. In early 1999, Healtheon, a leading US provider of health care Electronic Business applications, agreed to merge with a rival, WebMD, creating the largest such portal site for physicians in the United States.

More than 65,000 doctors in 11,000 practices can access a wide range of information and features via Healtheon's physicians' portal. The new service allows GPs to perform various business and financial transactions, such as checking eligibility, billing and ordering lab tests and prescriptions, via the portal.

Advertising and the Media

Selling digital information or "content" is another potential way to make money on the Internet. In the early days of the Internet, the industry's favorite phrase was "Content is king." It was believed that if you filled a Web site with enough high-quality content, people would pay to visit it.

More than 90 percent of Internet users are primarily looking for news or information, but few want to pay for it. This is the principal problem for any business hoping to sell content online.

The publishing industry recognized early that it had much to gain—and perhaps also lose—by transferring traditional paper-based content to the Internet. Newspapers were among the first to spot the potential for the Internet in reaching a wider audience than physical distribution would allow. Today, nearly every major national European newspaper has an online presence and many local papers do as well.

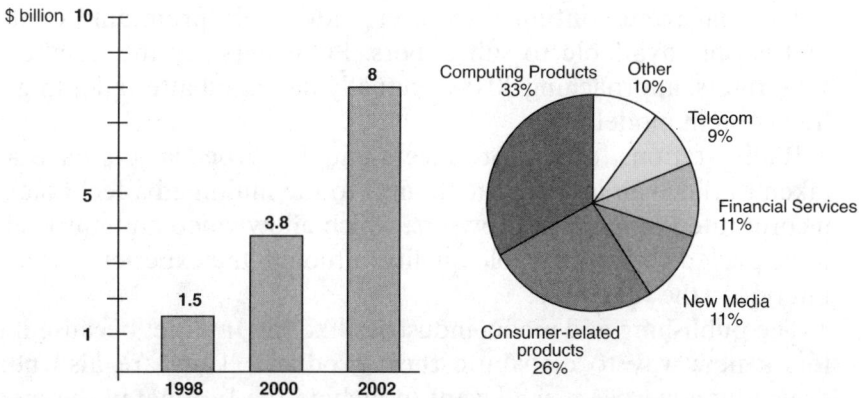

Figure 16. Advertising revenues on the Internet (IAB)

Some newspaper editors put up a small selection of each day's stories, or only allow full online access to subscribers of the paper product. Other editors give online users much more than their traditional readers, including multimedia, searchable archives and hyperlinks for users to visit the Web sites of companies mentioned in a story. A more recent development is to offer a rolling news services in which stories are posted on the Web site as they develop—up to 24 hours before they appear in the print version.

Most newspapers do not charge to read their online versions. Instead, they generate revenue through online advertising. Advertisers in the print version often get a "banner ad" on the Web page for marginal extra cost. One big-circulation newspaper that goes against the tide by charging to read its online version is *The Wall Street Journal*. Another is *The Economist*, which looks and reads like a magazine but calls itself a newspaper.

For specialist trade and scientific publishers, the subscription model can be attractive: The costs of publishing an electronic magazine accessible through a subscription-only Web site can be less than printing and distributing traditional paper publications.

Slate, an online magazine of US news, politics and culture backed by Microsoft, is one of the more famous experiments in online publishing. It provides a good demonstration of how different business models can be employed on the Internet. An Internet-only magazine available just to subscribers, in early 1999 *Slate* gave up charging a subscription and made the bulk of its content free. Over the following months, it experienced record traffic increases and was able to offer advertisers a larger audience.

The magazine continues to offer additional "premium" content that is only available to subscribers. Publishers say the number of subscribers, approaching 30,000, actually increased after adopting the free-content model.

Radio stations, record producers and TV broadcasters have also taken to the Web in a big way thanks to the multimedia technologies incorporated in modern browsers, which allow video and music clips to be played with reasonable quality, although the experience is often inferior to the real thing.

The publishing and media industries like the Internet because it offers a new way to distribute their products. They are also being driven by advertisers, who want to include the Internet in the media "mix" for their campaigns.

Unilever, the Anglo-Dutch consumer products group, is one of the world's biggest spenders on advertising. It plans to use 10 percent of its advertising budget for new forms of communication such as the Internet.

While still tiny compared to traditional advertising media, the Internet advertising market has grown dramatically over the past four years. In 1998, spending on Internet advertising doubled to $1.9 billion from $907 million a year earlier, according to the Internet Advertising Bureau, a New York–based association.

According to the IAB's figures, consumer-related advertising was the largest category in 1998, with a 29-percent share of the total spent. Computer products were second, with 20 percent, and financial services third, with 19 percent.

The growth of consumer-oriented online advertising is perhaps the most conspicuous development in recent years. It counters early critics who said that the Internet would appeal primarily to buyers of computer hardware and software.

Selling Sex ... and Gambling

Putting content on the Internet does not guarantee that a business will succeed. Even sex sites, of which there are at least 100,000, have a hard time standing out in a crowded marketplace. The sex industry may account for at least 10 percent of the total online retailing market—accurate figures are impossible to obtain—but its contribution is largely ignored.

Sex sites pioneered some of the most sophisticated Internet technologies and we are not just referring to those used to download hard-core video clips. The sex industry helped develop clever techniques for advertising, user-tracking and e-commerce that are now commonplace on more mainstream sites.

The pornography industry was the first to face a problem that is now common to all businesses hoping to make money on the Internet: how to distinguish itself from its competitors in a crowded market. Up to 30 million individuals a day log on to the Internet in a personal quest for pornography. Most limit their surfing to the 70 percent or 80 percent of adult sites that have free images. These free sites are used as "bait" for the subscription sites and make their money by successfully getting visitors to sign up for "premium"—meaning paying—services.

Selling advertising on popular Web sites might seem an easy way to make money on the Internet. But today's surfers are increasingly resistant to the "click-me" ads and advertisers are demanding more sophisticated techniques to better exploit this promising new medium.

Conventional Web sites can still get away with charging advertisers for the number of banner ads shown to viewers—this is called the number of "impressions." In sharp contrast, the free pornography sites, which are awash with click-me banner ads, only make money when a user visits an advertiser's site and makes a purchase.

Around 5 percent of the visitors to a free pornography site might click on an ad—this is known as a "click-through." One or two out of every thousand will purchase the services advertised, in a process known as "conversion." In an attempt to increase conversion rates, the online sex industry has had to develop some highly creative techniques for luring sex surfers onto the paying sites.

The newest and potentially one of the most lucrative areas of Internet service is online gambling. A new service will allow anyone to participate—where legally allowed—in the lottery programs offered in many US states and elsewhere around the world. In addition, there are many games that are being developed for Internet aficionados to play and gamble on the outcome.

While online pornography and gambling are unknown worlds to many people, the issues and challenges facing them are remarkably similar to those facing more conventional Electronic Business sectors. For instance, every Web site knows that building brand awareness through advertising and marketing is critical to success in a new and rapidly evolving market. Becoming a portal or an Internet commerce hub is all the rage.

Hitting the right demographic in any medium is key for advertisers. But unlike other media, Web-based technology allows for the tracking of readers. This is moving mainstream advertisers to copy from the pornography industry and spend their money on deals in which part or all of an ad's cost is based on click-throughs, or the number of readers that are driven to the advertiser's site. Click-throughs can give advertisers reader information that can generate sales leads, helping them gauge Web usage and buying habits.

This trend has been accompanied by an inexorable decline in online advertising rates that depend on simple "cost-per-thousand" measures. When Internet advertising started, it seemed natural to attempt to apply the publishing industry's standard techniques for selling advertising such as cost-per-thousand.

According to the Internet marketing company AdKnowledge, average cost-per-thousand rates across all types of online publishing fell 6 percent during 1998. The drop was steepest for online consumer magazines, where rates fell 24 percent. In professional business-to-business publishing, the decline in cost-per-thousand rates was 15 percent.

The rates are falling simply because the growth in the number of Web sites supported by advertising exceeds the growth in demand. The figures above are for US Web sites and in many European countries, Internet advertising is a much more recent phenomenon. This means that there are probably still opportunities in Europe to create a profitable Web business by selling advertising, but sooner or later the laws of supply and demand will catch up.

Conclusion

The Electronic Business revolution is affecting everything and the US business press is reporting each development in exhaustive detail. For example, *The Wall Street Journal* of May 10, 1999, had no fewer than 15 significant stories on the Internet and related companies, including major ones on the editorial page. Some examples:

- In a story on BankOne Corp., the fourth-largest US bank, CEO John McCoy, one of the most aggressive acquirers in the banking industry, said he was not planning any more major purchases but would instead pin the bank's growth on the Internet. "Only the largest banks will have the clout to compete in the borderless world of the Internet," he said.
- The editors of *The New Economy* were quoted as saying that "AT&T (is) already a dinosaur," in commenting on Microsoft's $5 billion investment in AT&T and the two firms' joint positioning for the Internet market.
- In an article headed " ... Or the Wave of the Future," two professors, one from Harvard Business School and the other from MIT's Sloan School, said, "At stake is the future of the Internet itself. Regulating on Internet time has to become like competing on Internet time!"
- Another headline read, "Merrill's (Lynch): Upgrades to Benefits Web Site to Include Help From Economics Whiz."
- A story on Net.Bank Inc. reported that this Internet bank plans to raise more than $165 million through an IPO.
- Another item was titled "Sports Internet to Buy UK Sport Web Site, Focus on Gambling."
- The article "CommonPlaces' Web Strategy Targets Students— CollegeBytes.com Site Aims to be the Internet Hub Among Undergraduates" highlighted another project aimed at the young.
- Almost all the major advertisements were related to the Internet.

This random selection of stories, taken on a typical day, demonstrates that the Internet and Electronic Business top the agenda of the premier financial publication in the United States. The message is clear: If you want to be competitive in the new economy, then the Internet and Electronic Business ought to top your agenda, too.

Finally comes the question of competition. On the same day, *The Wall Street Journal* reported on a conference in Boston, Massachusetts, at which executives at major companies considered whether they would like to be the "Deller" or the "Dellee," referring to how Dell Computer has outmaneuvered competitors such as Compaq in the PC industry. They concluded that Compaq is suffering partly because it has been "Delled"—outflanked by faster, cheaper products from a thoroughly digitized Dell.

European companies, be warned: Do you want to be "Delled"?

That message holds true whether you are selling advertising, nuts and bolts, or travel online. Particularly in Europe's less-developed regions or the more specialized markets or industries, the opportunities are there for the taking—today. It is only as matter of time before competition heats up. If the short history of the Internet has shown us anything, it is that those who establish an early lead tend to hang on to it.

The remainder of this book will explore these themes and explain how to build that all-important lead. But first some background on why Electronic Business is happening and why it is happening now.

2 Electronic Business Background

In this chapter, we discuss the external market forces that lay the foundation for Electronic Business. The major factors are the transition from geographic to electronic communities, globalization, the increasingly important role of the individual and the telecommunications revolution that is sweeping the globe. Moreover, powerful forces within organizations are creating the right conditions for the success of Electronic Business. Such internal factors include the adoption of federated structures, massive productivity gains due to the application of IT to business processes, the catalytic role of entrepreneurs in the new business environment and the recognition of the value of intellectual property rights as a business asset in the digital economy.

Transition from Geographic to Electronic Communities

Throughout history, the structure of human activity has been geographic. This applies at the local, national and global levels. Urban communities differ from suburban, which differ from rural. There are differences between people inhabiting coastal communities and interior communities. There are differences between ports and manufacturing centers. There are differences between cities and towns on the plains and those on the hills. Even in Silicon Valley, there is a difference between the "hill people," such as those who live in Portola Valley and Saratoga, and "flatlanders," who live in places such as Palo Alto and Mountain View.

Some of these differences are trivial; others are fundamental. Basic issues such as construction, the environment and traffic often pit geographic communities against one another; urban communities adopt a different attitude toward construction than rural communities.

Historically, most people did not have the freedom to choose the community in which they lived. But starting with the Industrial Revolution, people increasingly migrated to other geographic areas. In recent years, the ease of transportation has accelerated these migratory patterns.

Knowledge itself has often been geographically localized. The City of London, for example, became a repository of knowledge about finance and financial dealings; Amsterdam became a center of the diamond industry; and Silicon—now Software—Valley in California became a center of knowledge for the semiconductor and the software industry.

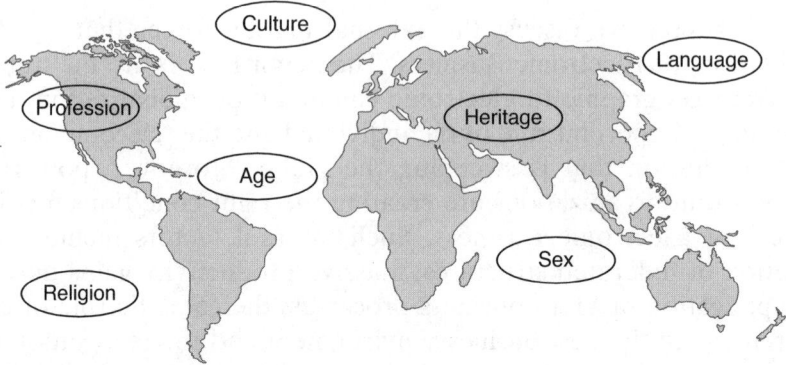

Figure 1. Electronic communities for individuals

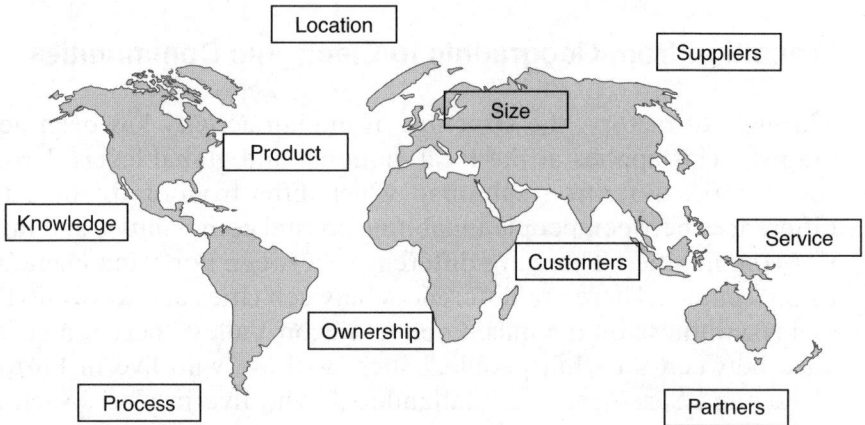

Figure 2. Electronic communities for businesses

In the Electronic Business world, the structure of business will be knowledge-based. Knowledge will be coded in digital form, collected and organized. It will then be portable. As a result, centers built on control of specialized knowledge will lose their importance to new centers better equipped to develop and employ that knowledge. The location of such centers may not be determined by geographic considerations alone.

Examples of non-geographic structures already exist in the consulting, legal and accounting professions. Many of the larger organizations in these fields have developed "practices," which are groups of individuals with expertise in a particular area and computerized knowledge databases. These are distributed around the world. Arthur Andersen and Andersen Consulting are examples of organizations that already embody this principle.

Geographic Boundaries Are Changing Rapidly

It seems that the second law of thermodynamics–the tendency to move toward disorder–applies to human activity as well as scientific activity. Our emotions and cultural tendencies seem to move us toward fragmentation at the same time that logic would impel us toward aggregation.

Thus at the same time that supranational organizations, such as the North American Free Trade Association (NAFTA), the European Union and ASEAN, are forming around a set of common interests, there are numerous examples of countries that are breaking up into component parts. Britain is evolving from a centralized state to a federated state, including Scotland and Wales as entities federated with England. Spain could well break into multiple parts. In Italy there is a potential divide between the North and South. And what is happening in the former Yugoslavia is an extreme case of fragmentation along ethnic rather than geographic lines.

One reason for this is that many boundaries in the world today are neither logical nor based on culture, but are the result of geopolitical forces such as historic battles, invasions and empires. Exceptions to this rule where boundaries are logical and relate to controlled cultures are almost always islands such as Japan. Thus there are two countervailing tendencies of aggregation on the one hand, often built around standards of doing business or defense, and fragmentation on the other hand, built around ever smaller cultural affinities.

Businesses Are Built to Serve Communities

Business must adapt to these tendencies of aggregation and fragmentation. Electronic Business allows companies to adapt to both of these trends simultaneously. Historically, we have had either multi-national companies, which conducted their business across national boundaries, or we have had small organizations serving local communities. Electronic Business makes it possible for local organizations to spread their cultural affinity worldwide by means of the Internet.

At the same time, large multinational companies can adapt their processes and business offerings to local communities in a way never before possible. For instance, a Hong Kong Bank can become a "local" bank for a Chinese person sent to live in Nice, France.

Not only does Electronic Business serve communities, it creates them. The equivalent situation has happened on a geographic basis in the past. The shipping industry for example created seaports. Air travel has created some of the largest 20th century "ports"—around airports. A large airport may employ as many as 50,000 people, larger than most towns. And the characteristics of an agricultural community next to an airport are different from the airport community itself. The agricultural community will be much more like an agricultural community in another country than it will be like the airport community. Certainly they will share some common interest, such as the weather, but many of the characteristics of their "business lives" will be different.

Telecommunications creates the possibility of electronic communities that will eventually become stronger than geographic communities. These communities could well be built around culture and heritage. One consequence of this is that people may eventually vote in communities of choice rather than communities of location. Some of the first indications of this can be seen in the growth of the power of absentee ballots in US elections.

Globalization

The extremely rapid transfer of ideas and fashions from one location to another around the world is one aspect of globalization. It is increasingly difficult to buy different products in one location from another. The fact that we as a species tend to appreciate the same products accentuates the similarity of humans rather than the differences. It doesn't matter how much Coca-Cola advertises, it would not achieve its phenomenal global market penetration if people basically did not like the taste. Advertising might persuade you to buy the first can of Coke, but the second has to be bought based on appreciation of the product.

The globalization of the communications and travel industries, which enables us to see, touch, feel, smell and taste products and services in other communities, gradually shifts our purchasing toward products and services with universal appeal. This trend also forces the elimination of lower-quality products and services that cannot compete. Governments may erect barriers against higher-quality products and services from outside their local community, but in the long run these barriers will be overcome by the wishes of the population.

Thus globalization involves breaking down geographic barriers. It can create new barriers such as those created by access (or lack of access) to telecommunications.

Improved communications therefore drives global competition. As we shall see, standardization of processes and offerings from vendors in the Electronic Business space will make global access to markets available to even the smallest organization. Historically, such global access was reserved for only the very largest multinational organizations, which could afford the physical infrastructure to support such operations.

In the 21st century, this aspect of Electronic Business will create massive shifts in the competitive structure of industries. Factors such as low-cost, high-quality, high touch-and-feel customer service will increase in importance on a global scale. Moreover, the ability of a supplier to adapt to the culture and the characteristics of individual customers no matter where they are will determine success or failure.

Global competitors will need global Electronic Business systems that will not only support the manufacturing and distribution of goods but also the service component of our economies. Today, only 10 percent to 15 percent of all services are traded across national boundaries, as compared with more than 50 percent of world production in the manufacturing environment. Electronic Business applied to the services industry can move the proportion from 10 percent to 15 percent to perhaps 40 percent to 50 percent within a relatively short period of time.

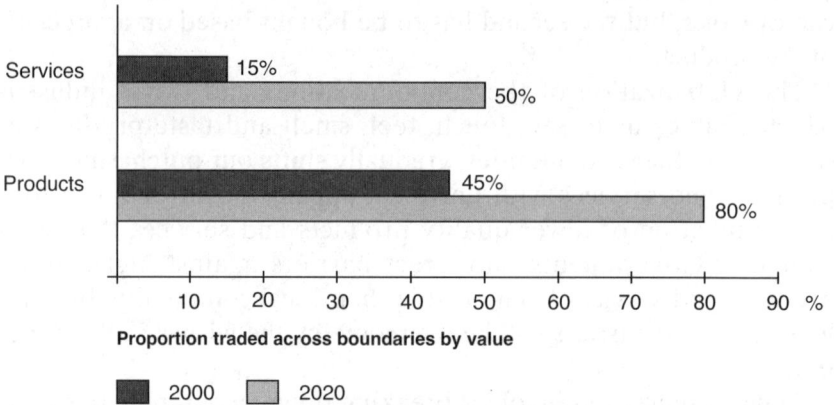

Figure 3. Impact of Electronic Business on world trade of products and services in 2000 and 2020 (INPUT)

Information and entertainment are certainly among the first services industry sectors to be affected by this Electronic Business globalization. For example, CNN is distributed globally, as is much programming from the BBC. Sports and entertainment now address global markets. Everyone knows Michael Jordan. Electronic Business aspects of broadcasting will accentuate and accelerate this development. In the 21st century, it will be possible to find, pay for and watch a soccer game between Juventus, Turin and Manchester United from anywhere in the world.

We will be able to watch shows, opera and theater wherever we are. Often it is these recreational, sporting and cultural aspects of our lives that define the communities to which we belong.

Telecommunications is not the only force driving the Electronic Business revolution. Containerization, for example, has had a huge impact on the globalization of industry. The ability to make something in Hong Kong today and have it in Frankfurt the day after tomorrow by loading a container on a Boeing 747 has transformed our manufacturing and retailing processes.

Package services companies such as DHL and Federal Express have both benefited enormously from and facilitated the shift to globalization.

This ability to shift goods quickly around the world has had a dramatic impact on the demand for supporting systems. As recently as 20 years ago—when it took seven days or more to turn around a ship in a harbor—information systems that supported the shipping process could take their time. There was time to interface with the shipper, with customs, the transportation agency, etc. Containerization,

with its 24-hour ship turnaround, radically redefined the system requirements. In the early days of containerization, a ship could leave
San Francisco and arrive in Tokyo without the authorities and companies in Tokyo knowing what was on the ship when it arrived. The
information systems were slower than the movement of the ship itself. Needless to say, this gap did not last long.

This story points out the importance of matching systems to business processes. These systems are not just information systems; these
systems include all aspects of the processes that enable them to operate, including people, organization and money. The drive to the
globalization of business and attendant systems has caused many
parts of these processes to be coded. Knowledge, which used to be
resident in the minds of experts, has been captured and coded. Once
coded it can be moved. One consequence of this is that a manufacturing company can set up a factory almost anywhere in the world in
a very short period of time since all the processes involved in establishing and operating a factory to produce products have been captured and coded.

This coding process is now happening in the services world. Much
of it relates to the development of the discipline of business process
engineering in the 1990s. Some people have termed it "business
process re-engineering." But most business processes were never
engineered; they simply grew.

It is common to think that business process re-engineering was a
fad of the 1990s that is now over. Nothing could be further from the
truth. Every major company and organization today is involved in
the engineering of its processes. One interesting discovery is that
processes are often far more common across industries than was previously imagined. In the past, there was nothing a bank in Milan,
Italy, could offer a manufacturing company in Milwaukee in the
United States. Business process engineering, however, has allowed
companies to look for the best-of-breed activity regardless of location
or industry. Now, a bank in Milan that develops an advance to customer manager systems does indeed have something to offer a Milwaukee manufacturing company.

Organizations such as Siemens Business Services and Arthur Andersen are capturing best-of-breed processes so they can be transferred across organizations.

Traditional   **Future**

Size of company (annual revenues)   Number and type of customers

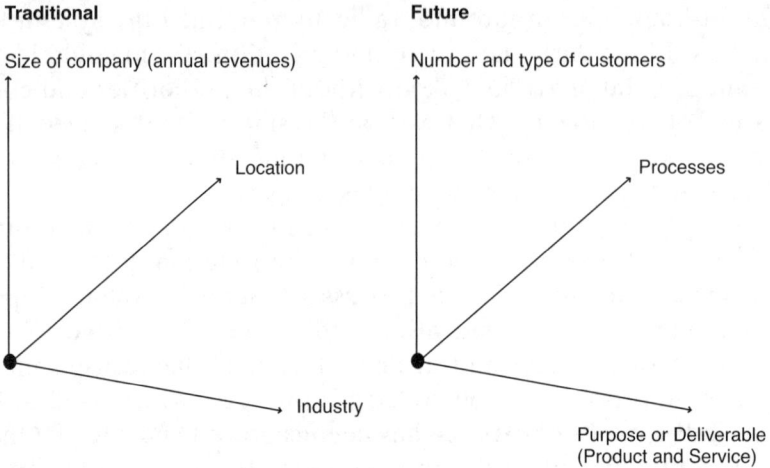

Figure 4. Shift in organizational definition criteria

These developments will increase the competitive pressure on lo-cal organizations. In many cases, the last remaining inhibitor to glob-alization will be government or quasi-government regulation that protects local businesses. Examples of such protection are accounting rules, employment regulations or other local regulations, such as data privacy, that can be particularly important in this regard. In the 21st century, protected markets such as telecommunications, airlines, utilities, banking and insurance will quickly lose protection.

Globalization of Ownership

Globalization of ownership will drive the creation of a truly global market. Opportunities for international ownership are increasing despite restrictions that often hinder individuals in their ability to directly participate in the ownership of organizations abroad. IT sys-tems are surmounting barriers erected by governments to protect themselves and their institutions. It is now possible to purchase equi-ties and other instruments around the world from virtually any coun-try through systems such as those developed by Atos, the French information services company. The Fimatex Global Trading System from Atos allows investors anywhere to place orders on a variety of stock exchanges.

In addition, financial and company information is easily available in both print and electronic form almost anywhere. As a result, indi-viduals will invest in companies and organizations outside their geographic boundaries. This is the beginning of a global equity market. It already exists at the institutional level. Countless funds are devoted to investments on a global, national or local scale. This is a well-accepted vehicle for individuals to participate in investments in countries other than their own.

Electronic Business provides a fundamentally different capability: It enables an organization or an individual to make purchases directly without having to go through an institution.

The corollary to this is that companies can make stock offerings without having to go through a broker. The first companies in the United States have gone public using the direct sales method. The SEC has explicitly condoned this process under properly controlled conditions.

This is electronic investing. Today's electronic brokerage activities from Charles Schwab, E*Trade and others are merely steps toward a truly electronic investment market rather than being true electronic investment systems themselves.

This movement will create the need for new services, particularly those associated with certification and authentication. Online escrow, clearing and trust services will also become extremely important.

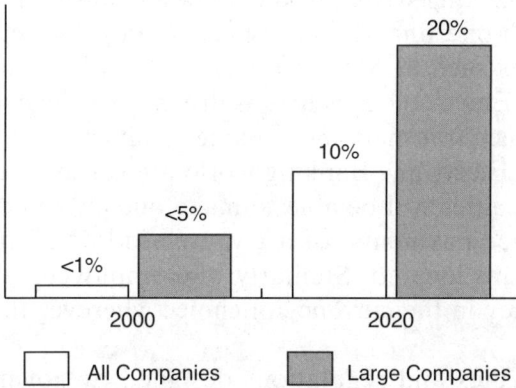

Figure 5. Globalization of ownership of companies in 2000 and 2020

The acquisition process is also a factor in the globalization of ownership. When a German company such as Deutsche Bank acquires an American company, ownership of the resulting organization is often spread among American and German investors. Thus, the acceleration of international acquisitions steps up the globalization of ownership.

These two factors, the ability to invest globally and the corporate distribution of ownership internationally, will mean that organizations will gradually lose their national affinities.

Their corporate governing bodies likewise will shed their homogenous nature as boards of directors become multinational. This trend started in the United States, but has since spread to Japanese and European organizations.

Thus the globalization of business will be accompanied by the globalization of ownership and governance, aided and supported by Electronic Business.

Globalization of Financial Systems

People often comment on the power of the global financial system. They point to the trillions of dollars that are traded through the SWIFT electronic funds transfer system.

But this is not a true global financial system. It is merely a global financial trading system. Money is traded among what are essentially local institutions; banks, mutual funds, investment companies, etc. Even multinationals are usually still located in one geographic place. A truly global electronic banking system would allow an individual or a corporation to deal with one financial entity across all geographic boundaries. Organizations such as Merrill Lynch and CitiBank are striving toward this goal. Electronic Business is driving the financial services industry toward such transnational banking facilities.

The possibilities of the Electronic Banking world are certainly enticing. For example, companies will be able to make one payroll deposit to cover all employee payments for a given period no matter where those employees are located. Similarly, the employees will enjoy access to their money in the currency of choice wherever they happen to be.

Today, local banking rules and regulations designed by governments in order to protect their local banks make this impossible. But these rules create inefficiencies that impede the growth of productive organizations. As such, they will disappear in the Electronic Business world.

It is now apparent that investment attitudes are changing. The Internet phenomenon is attracting investment from all parts of the world. The stock prices of many embryonic companies in this electronic world are so high because people want to be part of this revolution.

Recently, a lawyer addressed a group of executives of such companies by saying, "You are the Henry Fords of the 21st century. I don't know exactly what you do. All I know, is that I want to be part of it."

Thus billions of dollars are being invested in the infrastructure, enabling tools and services to support the growth of Electronic Business. In 1998, more than $10 billion of venture capital was invested in technology in the United States, according to a survey carried out by PricewaterhouseCoopers. Of this amount, investments in Internet companies totaled $3.5 billion, up 66 percent from the previous year. These amounts could reach $15 billion by the early 2000s; this is a staggering amount of investment. Much of it, of course, is recycled; people and companies that have made huge profits on early investments are reinvesting. This "pyramid scheme" will continue until there is a substantial retrenchment.

Established major organizations are now spending equivalent sums of money on Electronic Business. But it was a combination of individuals whose minds were open to this new world and entrepreneurs that created this revolution. Established organizations and governments, with the notable exception of the US federal government, have generally been latecomers in this investment and development process.

From the Organizational to Individual Mindset

Since 1989, there has been global acceptance of the concept that the "free" market economy has defeated the centralist, communist economy. This was not the real conflict. In many respects, the market economies are as strictly or even more strictly regulated than the communist economies.

The fundamental difference between the two is the concept of individual ownership. In a communist society, individual ownership of property does not exist. In the "market economy," the concept of individual ownership is paramount. Note that it is possible to have a market economy in which ownership is not by the individual, but by organizations or large entities.

Thus the success of the market economy model is due to the success of the concept of individual ownership versus group ownership. This represents the general trend. In all aspects of human affairs, the individual is winning. The individual is winning in government, in business and in society. We speak of a free society meaning that the individual is free, not the organization.

In the Electronic Business world, we have the ultimate of individu-ality wherein individuals own their identity. They own their domain name or names. In fact, many individuals have registered their name as a domain. These domain names attain real value. Many of them will be handed down from generation to generation through wills.

Electronic Business fits perfectly into this concept of the primacy of the individual. It shifts the power from the seller to the buyer. It shifts power away from the provider to the recipient.

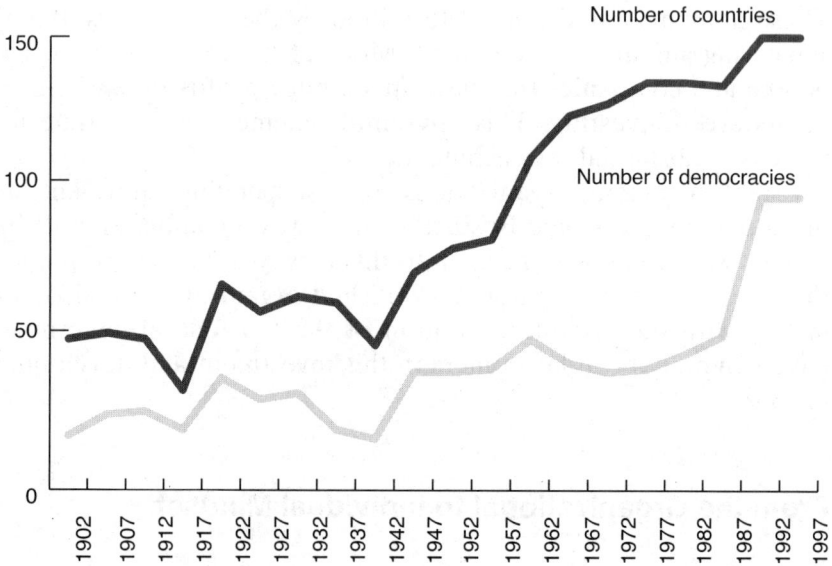

Figure 6. The increasing number of independent governments worldwide

This does not mean that the rights of the group are totally ignored. For example, the good of the group, defined somehow in terms other than a national group, demands individual conformity. The contro-versy about the extent of permitted pornography on the Internet is an example of the struggle between group and individual rights.

Electronic Business does not represent complete anarchy, as some would like. It actually represents the federation concept, in which the rights of the individual and those of the group are considered to-gether.

But the Electronic Business concept does affect the definition of the group. For the first time in centuries, religious affinities are be-coming more significant than national bonds. The concern of Muslim authorities over certain aspects of the Internet, for example, tran-scends national boundaries.

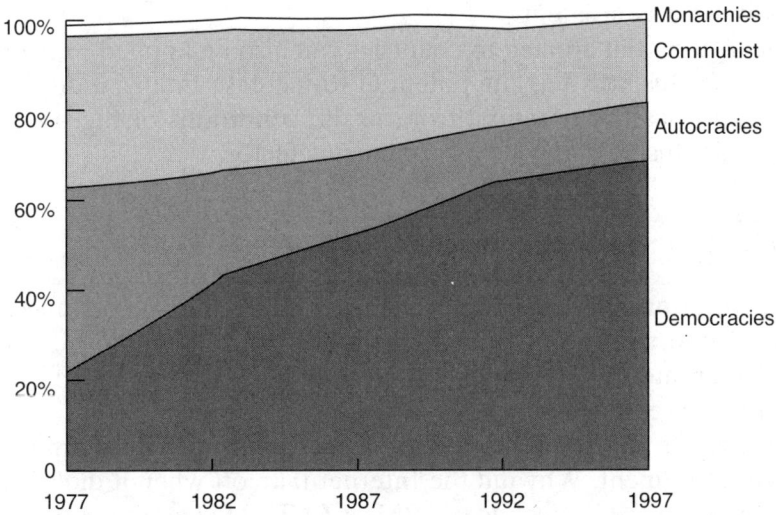

Figure 7. The percentage of the world's population under various political systems

The trend toward individualism has been most apparent in communist societies, but they are not alone. The deregulation drive in Western societies set in motion by former British Prime Minister Margaret Thatcher and former US President Ronald Reagan has gained speed around the world. But Thatcher and Reagan were not so much the initiators of the message as its deliverers, just as physicist Albert Einstein was the deliverer of the relativity message. If they had not existed, others would have picked up those messages because the time was right.

The world is too complicated for central management of all activities. There are too many dependent variables. So some form of sub-optimization is better, often accomplished through a federated environment.

Today's developments are a continuation of a century-old trend in government toward liberty and democracy. At the dawn of the 20th century, there were only 12 democracies in the world. There are now more than 150. The days of kings and dictators are finished.

This is not to suggest that the global Electronic Business economy does not face serious challenges. It will have to address the increasing disparity in wealth between the various communities on Earth. In his book, *The Wealth and Poverty of Nations,* David Landis points out that the wealth disparity today between the richest and poorest countries may be as high as 400:1, whereas several hundred years ago it was about 5:1. The electronic world has the potential to reduce this disparity or accentuate it.

The differential applied to countries can also be applied to people. Electronic Business has the potential to increase financial disparity. But it also has the potential to raise the conditions of the poorest. Electronic Business may be a harmonizing factor.

Revolution in Telecommunications

The revolution in Electronic Business is dependent on advances in telecommunications. As we noted above, telecommunications changes create the possibility of electronic communities and businesses to serve them.

The growth of the Internet is the most important telecommunications development. Why did the Internet take off when it did? It certainly was not the brainchild of IBM, AT&T, BT, Deutsche Telekom or any other major company. Government sponsored its initial use to support university research and the sharing of information. Yes, there were certain initial defense aspects of the networks but these quickly became subservient to the research objectives. Its transfer into the business community did not come from big companies, organizations or governments, but from individuals. People left universities and took the Internet with them. Initially they used it for research purposes, then applied it to the business aspects of their organizations. They also took it into their personal lives and interactions. The earliest uses of the Internet were for information transfer and messaging. It was also used for collaboration and for games.

The Internet took off because it appealed to the fundamental human need to communicate. It is the same reason that postal services have exploded over the past 150 years.

Moreover, the Internet took telecommunications power to the people in the same way that the personal computer took computing power to the people. In both cases, the establishment resisted this revolution. Bill Gates himself, who now embraces the Internet, earlier tried to kill it. In fact one can admire the ability of Gates and Microsoft to change direction. At one point, they were actually fighting the Internet at every step of the way and threatening that MSN (Microsoft Network) would overwhelm it. And then six months later they announced that MSN was part of the Internet and they were embracing it. There are few organizations in the world that are able to change with that degree of alacrity or, some might say, cynicism.

One appeal of the Internet is certainly its cost model. In the United States in particular, charges for the Internet were based on a monthly flat fee regardless of extent of use or distance between and among users. By demonstrating that time- and distance-dependent charging were anachronisms that telephone companies perpetuated in their own self-interest, the Internet is actually very much like the little boy in the story about the emperor without any clothes.

An electron appearing in Munich as part of a message does not have a little flag on it that says, "Hello, I'm an electron from Australia. I have come a long way. Please pay more for me." Yet telephone companies have conditioned us to expect to pay highly for international messages. This of course harks back to times when international messages had to be scheduled and routed through human operators. Today, the only incremental operating costs for distance- and time-dependent charges are those related to the collection of data for billing purposes.

With modern technology, telephone company charges for long distance—and international charges, in particular—result in virtually obscene profit margins. The cost of a transatlantic call today is less than one penny for calls on circuits from Global Crossing. Global Crossing was the company that laid a private optical fiber cable across the Atlantic and then sold its bandwidth to telephone distribution companies. With a pricing structure of less than a penny per minute per call, their profits will still be substantial.

Nature abhors two things: a vacuum and high gross profit margins. These high gross profit margins in telecommunications can only exist when governments protect them. The two pricing myths in telecommunications are distance-dependent operating charges and time-dependent operating charges. Once somebody is connected, the cost of operations is virtually nil, except for the opportunity cost of someone else not being able to use that circuit. This lost opportunity cost is also disappearing rapidly because of new technology that enables multiple users to share the same "pipe."

We as consumers and businesses have accepted these charges because of our physical world analogs. It costs us more money to travel a long distance and it makes sense that we pay more money the longer we use something. As we pointed out earlier in our discussion of virtual malls, physical analogs do not always apply in the electronic world.

It was the Internet that really opened our eyes to these telecommunications company charges. As voice shifts to VUD (voice under data) and becomes part of the digital stream of information, pricing systems will migrate toward low-price, flat-fee Internet charging systems. This will open up the use of the Internet to a wider community.

Internal Processes

Organizations as Federations

Earlier, we introduced the concept of the power of the federated structure. It allows for the centralization of certain activities, if not in terms of operations at least in terms of governance and regulation. It also allows for the recognition of the rights and desires of individuals and local communities. In this context, the word "local" might apply to communities in a cultural as opposed to geographic sense.

The world is too complicated and it's changing too quickly for one central organization to manage all activities. Our bodies apply the same principles of delegating certain decision-making processes away from the brain. When you touch a hot plate, you withdraw your finger before your brain has had time to process all the information and make a decision as to the benefit in doing it.

Organizations, particularly business organizations, are no longer rigid, hierarchical, wholly-owned structures. Modern organizational discussions tend to focus on a network of stakeholders who are owners, employees, managers, suppliers, customers and partners. They include full-time workers, part-time workers, temporary workers and contractors at various levels. The organization may well be geographically dispersed. This book, for example, is a collaboration of writers in three locations. Increasingly, cross-ownership of new entities is being established by partners in international business. Sometimes, these cross-ownerships span industry as well as national lines.

In all this, there is recognition that the customer does not always buy, or indeed want to buy, the component parts of a solution. The customer wishes to buy a solution, which is often best provided by a team of suppliers. Electronic Business is accentuating this phenomenon through its ability to network previously separate people and companies. Both vendors and suppliers recognize that no one company has all the answers. And the customer has neither the interest nor the time and energy to put together a team to provide the solution. In a sense, the customer is "outsourcing" the team-building activity. The customer expects the suppliers to build the teams to provide the solution.

The customer does not always expect to see the same team in the market, but does expect the team to hold together during the period of its relationship. Thus such teams are far more than "one-night stands" set up to address each opportunity on an ad hoc basis.

This concept of teaming on a contract-by-contract basis has been adopted in US government contracting without a great deal of success. Teams in the Electronic Business environment must be relatively close-knit and exist for long periods. These will be like a soccer or rugby team in contrast to a tennis or golf team. Tennis and golf teams tend to be assemblies of individuals who are performing best at a particular point in time. Soccer and rugby teams are an assembly of individuals who train and work together to accomplish their objectives as a team rather than as individuals.

Dictators Are Out in Organizations as Well

Just as we have observed in the political world, the concept of democratization has spread through business organizations. Ironically, workers unions have not played a significant role in this process. In most developed countries, unions have become less powerful. The fundamental reason is that the unions were also in the business of minimizing the individuality of the worker and tried to treat workers as rigid, homogeneous groups.

There is a drive toward worker empowerment and, in many countries, worker ownership. This has been recognized in the European Union in a different way from in the United States. In the EU, worker representation has been left in the hands of the unions rather than in the hands of the individuals. In the United States, worker representation has been achieved far more through worker ownership, whereby employees own shares in the company in which they work.

Traditional Business Structure

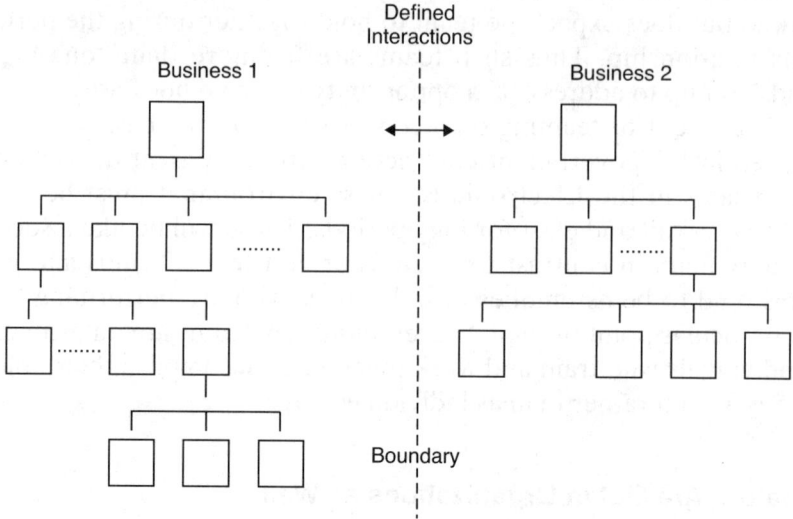

Electronic Business Structure

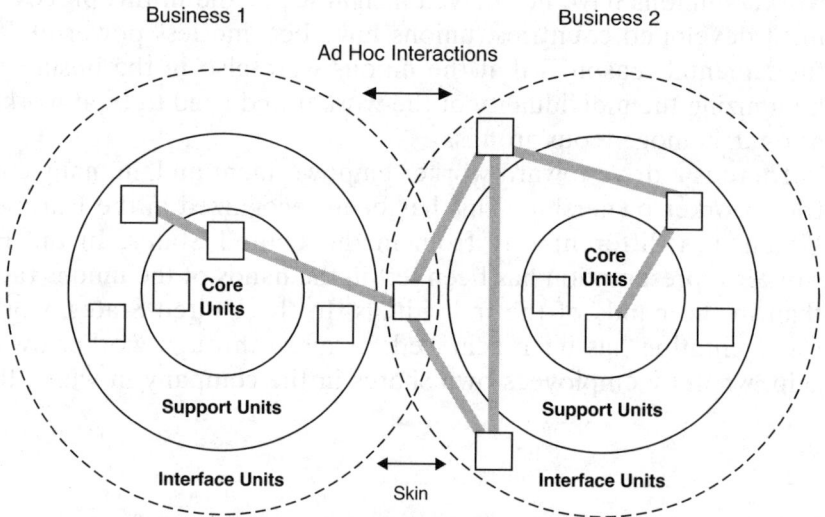

Figure 8. The interaction of business changes radically in the Electronic Business era and is characterized by project- and process-related teams made up of units from each organization's core, support and interface units.

Many have traced the hierarchical organizational structure of businesses to traditional military organizations. The Romans organized their soldiers into units of 10 arranged in units of 10, arranged in units of 10, and so on. Such structures were necessary to distribute communications to each soldier in battle conditions. They were also necessary for effective communications from the field to the head-quarters' organization. Communications today are revolutionizing military structures to provide more information and autonomy to local "managers" and at the same time, more central direction. The middle is being eliminated.

Instead of being at the top of an organization, planning and control units are now part of a core set of units surrounded by support units and then by external interface units.

The same system is applied to business structures in which quality of unit and individual performance is accentuated by flat organizations. The traditional span of control of 10, 11 or 12 people is giving way to spans of control of tens, if not hundreds of people, where the management and coordination role is supported by powerful collaboration, communication and analytical systems.

Communication is the most important and critical factor, and it is here that the Internet was first used for Electronic Business—to support and provide e-mail. This rudimentary form of communications provides point-to-point message transmission.

In order to support Electronic Business activities fully, however, multi-point to multi-point communications and collaborative systems capabilities are required. The Internet provides this. Thus electronic catalogs can provide information to all suppliers or all customers and prospects at the same time. Financial information can be placed on a Web site that is accessible to all investors at the same time, obviating the need to transmit individual messages. All that is necessary is for the target to know that the information is available when and if it is needed. In some cases, sending a flag that new information is available can be useful. The concept is that individuals or organizations come to the information when they need it, as opposed to sending information when it is available to a set of people regardless of whether they need it at that particular time. This is JITI (Just-In-Time Information)!

One of the consequences of the current approach is that individuals and organizations always place information into storage so that they can retrieve it when needed. This has resulted in a huge increase in computer and information storage capacity requirements. Collaborative approaches, including Web-based systems, dramatically reduce the need for storage, both centrally and locally, in most organizations. But there is still a human tendency to hold on to information just in case it may be unavailable at the central or producing site. It will take time to overcome this acquisitive tendency.

Productivity Improvements

The initial thrust of the US government in 1994 and 1995 for the information superhighway was to improve productivity. The theory was that productivity improvement creates wealth. The information superhighway (electronic society) would boost productivity and hence the wealth of society, particularly the US society. This reasoning led the United States to foster the Internet and what is now known as Electronic Business, but was then known as electronic commerce

Improved productivity either means lower prices or increased profits. Increased profits, at least to the extent enabled by Electronic Business, are transient if they are exorbitant. Other companies simply move in to the same business area.

Therefore, in general, improved productivity from the information superhighway has resulted in lower costs and lower prices. This has been a major inhibitor to the growth of inflation even in a time of relatively full employment in the United States.

The basic systems engineering of this concept originated in the late 1980s and early 1990s and was incorporated into business process engineering as described above. This process resulted in methods such as JIT (Just In Time), SCM (Supply Chain Management) and ERP (Enterprise Resource Planning), to name a few.

This productivity improvement process is by no means finished. The advances in computer and communications technology ensure that thresholds of choice in processes of all kinds are constantly changing. Although physical costs are rising (postage, for example), electronic costs are falling at ever increasing rates (telephone costs, for example).

It used to be an accepted measurement in the computer industry that there would be price-performance improvements each year of 20 percent. Those days are long gone. Today, we expect and receive annual price-performance improvements that can range as high as 50 percent to 60 percent in areas such as communications costs, electronic information storage costs and central processing costs. Such price-performance improvements will continue to have an impact on productivity, especially in the more IT-developed societies and industries.

We almost have a "virtuous" cycle. In Electronic Business, IT becomes more important. IT productivity improvements have greater impact. This, in turn, leads organizations to invest more in IT, resulting in increased productivity and so on.

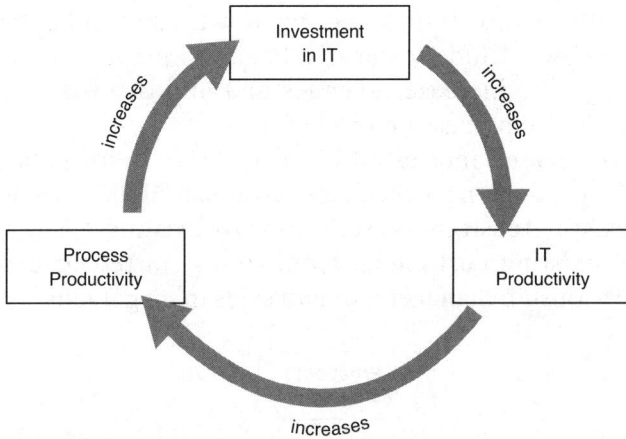

Figure 9. The "virtuous" cycle of IT productivity affecting process productivity

In summary, the higher the proportion of IT in a particular process, the greater the impact improvements in IT productivity will have on it. This is one reason why companies, regions and countries seem to be pulling ahead of others that, a few years ago, were regarded as being of equal or parallel capability. A good example in the retail industry of a company attaining a competitive advantage by investing heavily in IT is Wal-Mart.

The results are in. IT and Electronic Business works. In the first quarter of 1999, productivity in the United States increased by 4 percent. It is now generally recognized in the media and other areas that IT does produce improvements in productivity as well as enable the performance of new activities that otherwise would not be possible.

It is impossible to overestimate the value of positive attitude toward IT, the Internet and Electronic Business. Virtually nowhere is there any coherent, organized and established opposition to this shift, at least in the United States. This is unusual historically where there has normally been opposition to change from traditional institutions, including labor representation.

Today, there seem to be very few Luddites in the world. Socially, economically and culturally, there is a fertile platform from which Electronic Business can develop.

Productivity No Longer the Only Key

In the late 1980s and early 1990s in the United States, the emphasis was on reducing costs and improving productivity through methods such as business process re-engineering, downsizing and outsourcing. But in 1995 a new attitude toward IT began to prevail in the United States. The drive to increase revenues and improve front-end processes replaced this emphasis on productivity.

Companies became interested in using IT to help them get into new markets, produce new products, establish higher renewal business from existing customers, sell more to existing customers and better serve existing customers. Companies started to emphasize customer relationship management and sales management.

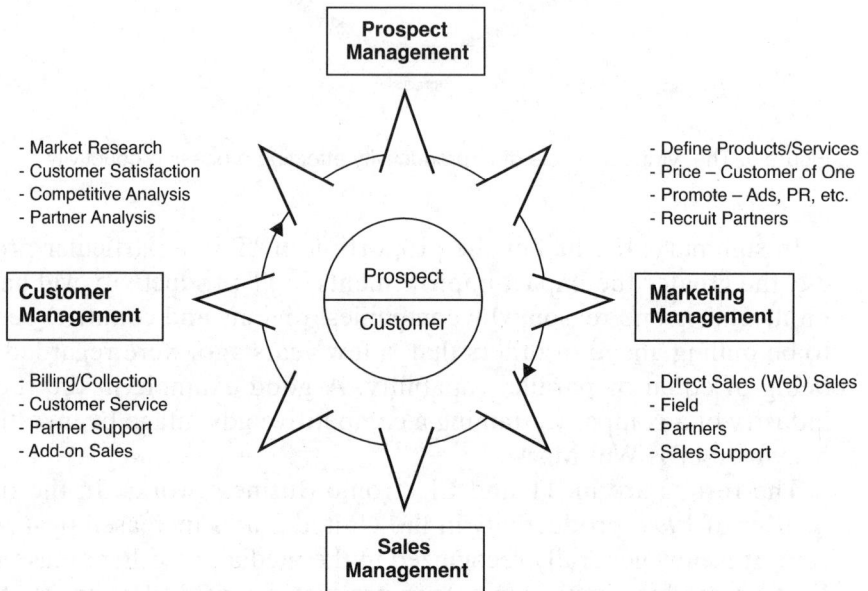

Figure 10. Customer- and prospect-focused Electronic Business processes

As a consequence, new software companies such as Siebel Systems grew very quickly while some of the more traditional ERP vendors started to slow in growth during the late 1990s. Companies such as Baan and SAP rapidly re-engineered their offerings to support the interest in new revenue generation systems.

All these externally-oriented processes required telecommunications networks. In the late 1990s they also required new people, structures and organizations. This explains the dramatic move toward tele-sales, call centers and electronic marketing. (It has been argued that the "800-number" phenomenon is truly the first example of technology applied to the electronic sales process.) In the next five years, call centers will switch from a primarily human/communications network environment to a computer/communications network environment. In the early days, call centers were simply collections of individuals, sitting with manuals and answering telephone calls. They are being replaced by highly sophisticated systems with artificial intelligence, business intelligence tools, sophisticated and expensive database and database handling systems, automated call balancing, call forwarding and response systems, voice processing systems, and much more. This is applying IT to improve productivity in the IT sector itself.

To date, much of the focus of these new systems has been on customer relationship management. In many cases, call centers have primarily been devoted to handling inbound calls for customer service. But companies such as Dell recognized early on that these calls could be translated into sales opportunities. Many call centers are switching over to outbound activities as well as handling inbound customer inquiries and sales support issues.

As companies move toward electronic sales and business, the emphasis switches from customer relationship management to sales and marketing management.

In Electronic Business, the buyer has the power. Companies that cannot reach the buyer lose. As a consequence, electronic marketing, followed by electronic sales, will dramatically increase in importance. The early signs are already visible. Advertising on US television networks has dropped sharply in the last year as advertisers switch their advertising dollars to the Internet and Internet-related vehicles. The very high value of portals such as Excite, @Home or Yahoo, is due to the marketing budgets that are being allocated to them.

Thus prospects become a very important concern. Each customer must be treated as a prospect. Many a wife or husband would rather be treated like a girlfriend or boyfriend; they would like to be constantly wooed. The same is true of customers. They must be constantly wooed in the future because switching costs (from supplier to supplier) will often be much lower than in the past and the ease of switching suppliers much greater. Electronic Business changes the relationships of suppliers, prospects and customers.

The Entrepreneur as Leader

To a large extent, the growth of the Electronic Business phenomenon, particularly the Internet, has been the story of entrepreneurs as opposed to large organizations. Electronic Business and its components have been dependent not on big companies, but on thousands of small software and service companies that over the years have applied technology to users' needs. These people have built big businesses from virtually nothing.

There are literally thousands of companies in the United States and around the world that are building the Electronic Business industry. Just as at the start of any major industrial change, it is the entrepreneurs who are providing the initiative. Many of these companies will fall by the wayside, but meanwhile they are exploring the boundaries of Electronic Business.

The two kinds of company involved are product companies and services companies. Product companies tend to be led by high-profile individuals. Services company leaders generally follow the maxim "Talk softly, but carry a big stick!"

Bill Gates of Microsoft is a classic example of a product company leader. Microsoft initially took computing power to the people and is now taking Electronic Business to both people and organizations. Other product company pioneers include Larry Ellison of Oracle, Scott McNeely of Sun Microsystems and John Chambers of Cisco. These companies have been extremely important in the developments that led to the Electronic Business revolution.

Executives in Electronic Business services companies with at least a billion dollars in revenues such as ACS, ADP, CSC, First Data, IBM and EDS in the United States and Atos, Cap Gemini, SBS and Sema in Europe are more self-effacing. But their companies have been instrumental in applying technologies from Gates', Ellison's and Chambers' companies to the users' needs.

The software industry leads the entrepreneurship trend. It relies on intelligence and people, rather than capital alone, and it is capable of rapid scaling. Is it such a surprise that investment money is pouring into these organizations? Every day Electronic Business companies go to the public stock market in the United States. Initially, these were only technology companies, but recently they have been companies that have been focusing on Electronic Business opportunities such as Web MD in the healthcare field.

The entrepreneurs founding the new companies are often from the new computer-literate generation. These are people who grew up with PCs and the Internet and are now in their early 30s. Some of the more innovative companies have been founded by people in their 20s and sometimes in their teens.

At the same time, global cottage industries are emerging. Teleworking is becoming commonplace. According to some estimates, as much as 20 percent of the working population in the United States works at least part of the time from home. Combine this with organizations that have people constantly traveling and there could be a major reduction in the bricks-and-mortar requirements for office infrastructure emerging over the next 20 years. In fact, Electronic Business is already impacting the real estate market. There is a perceptible trend to convert centrally located office space in cities such as London back to residential space, while companies move to new locations outside city centers and to networked units of small and home offices.

Intellectual Property Rights

The concept of intellectual property as a capital asset is critical to the development of Electronic Business. Assessing the value of intellectual property becomes necessary in acquisitions when the intellectual property rights (IPR) of a software or technology company form the main asset of the acquired organization. Other assets might include the customer base, distribution channels or sales force, but in the Electronic Business arena, these tend to pale in value relative to that of the intellectual property rights of the company's core products.

In the United States, there have been several acquisitions of companies that have zero revenues, let alone profits, for substantial sums of money, because of their perceived value in terms of intellectual property.

There have also been acquisitions where the main value was "eyeballs"—the number of potential viewers or surfers visiting a Web site. But, these eyeballs have no value unless they are captured, a process that involves collecting information on the identities and demographics of the people involved. Thus the eyeballs are converted into an IPR.

Recognition of the true worth of IPR is only just becoming prevalent. It is becoming an important trade issue among countries. Thus, the United States, where IPRs are highly valued, is conducting serious trade negotiations over software piracy, particularly with countries in Asia. In many of these countries, culturally speaking, information is regarded as being free. The Japanese have a saying: "Water, air and information are free." It is difficult to inculcate or import a Western view of the value of IPR into such a society.

Because IPR is a real business asset, it will be necessary to implement protective measures against piracy and sabotage at the individual, group, organizational or company, and national level. US military and government IT installations have already experienced attacks at the national level. The first reported coordinated attack by an outside agency occurred in 1999. Such attacks could spread beyond government installations to financial, retail, transportation or other targets, particular if they are seen as "soft."

Location is another IPR-related issue. Many IPRs are created in multiple countries and even on airlines between countries. Thus it is difficult to establish a geographic location for what is a non-physical product.

Also people and companies move their IPRs to areas of low taxation. Companies with IPRs in tax-free environments license them back to high-tax environments, where they only pay taxes on the value added in that geographic location. This problem will be compounded as business entities are increasingly established in offshore environments. The chief financial officer of Intel was recently quoted as saying before a Senate subcommittee that if he knew then what he knows now, he would never have established a company in the United States, but in the Cayman Islands.

As long as Electronic Business is relatively small, these impacts are minor. Since the exposure of organizations and companies to the possibilities of such an environment are just starting, few people understand its potential. But this will change dramatically as consultants, accountants, lawyers and others educate businesses in what is possible in the Electronic Business world. Of course, they will first have to educate themselves.

3 Electronic Business Technology

Over the past three decades, there has been a continuous flow of new technology in the fields of communications, consumer electronics and computing. In the last five years this has become an explosion in and around the Internet. The Internet is not a technology per se: it is a construct of technologies. The Internet technologies are the enablers of Electronic Business.

The Internet is often called "the network of networks." This concept of one integrated network having the capabilities and the capacities to include and/or carry many other diverse networks is the bedrock of what makes the Internet such a powerful force for change.

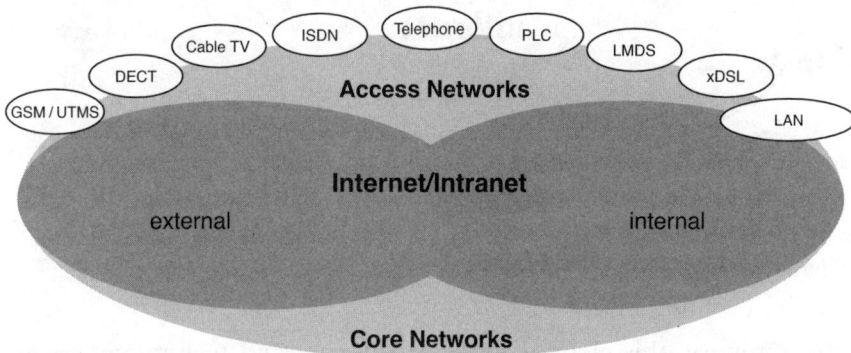

Figure 1. Internet—the network of networks

Beneath the Internet and surrounding it are numerous technologies that will be reviewed in this chapter as they affect Electronic Business.

Our emphasis is on software rather than hardware and communications, and on the ways in which businesses will adapt these technologies. We will examine how organizations make the transition from the older computing systems, dominated by mainframe and client/server structures, into the new structure of the Internet/Intranet. This has been dubbed "network computing" by some although it is hard to imagine what "non-network computing" would be like in future. Stand-alone systems will be as dead as the dodo!

We will discuss Intranets, which are the application of Internet tools, connectivity and networking within a business or organization. They are one of the fastest-growing segments of the Internet phenomenon.

The Java programming language and parallel changes in the nature of computing will be examined. These include the changes that are occurring as computing moves from the age of the PC (personal computer) to the age of the non-PC device, the Internet Access Device (IAD).

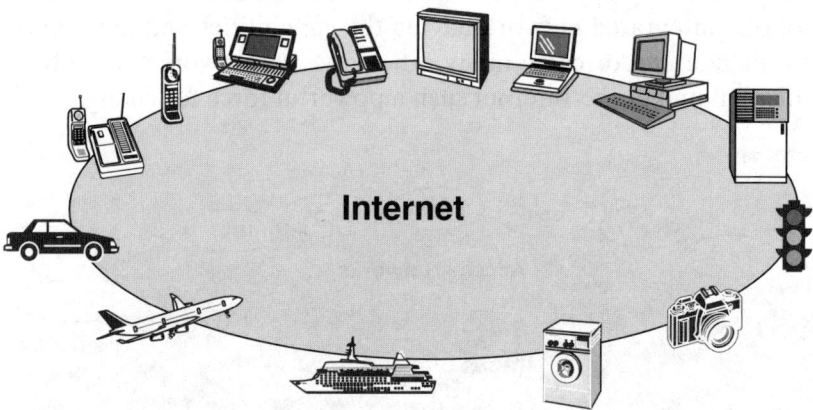

Figure 2. Device access to the Internet

In summary, this chapter looks at Internet and technology developments from the internal aspect of the Electronic Business revolution. The other sections of the book focus primarily on the external impacts of the Internet and Electronic Business.

Basic Technologies

The immediate future will bring three large rivers of technology together. The first river consists of computers, which have been digital for a long time. There were some analog computers in the '50s and '60s, but digital prevails today. The second river is that of telecommunications. All forms of telecommunication products and services are going digital on a worldwide basis. The third huge river is that of consumer electronics. Radio, stereo, TV, games, etc., have all been analog. These are very quickly going to digital. Even HDTV, which was proposed by the Japanese as a high-level analog service, has been overtaken by digital HDTV, as proposed by the United States.

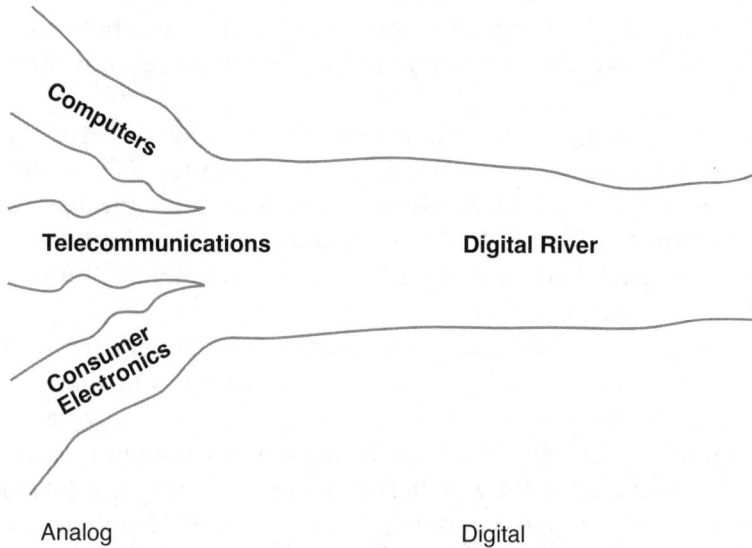

Computers

Telecommunications   **Digital River**

Consumer Electronics

Analog   Digital

Figure 3. Converging technologies—the technologies for computers, telecommunications and consumer electronics are converging into a digital river

In computers, it was the PC that took computing to the people, whether individuals were working on their own or as part of an organization. Telecommunications and consumer electronics have both been people-oriented from the start.

As these three rivers merge, we will see an enormous flow and current with blurred edges. The Internet is the channel for these rivers to flow together.

A multitude of other technologies has had and will have an impact on the Internet and Electronic Business. For example, price performance improvements in processors, particularly microprocessors, have made access to the Internet feasible. It has allowed for the growth in data, audio and digital interfaces with the networks.

Over the last several years, price performance improvements in processors have far outstripped those in communications. So much so that blame for poor Internet access, speeds, times and services has fallen almost continuously on the telecommunications companies.

Over the next few years, this may change in many geographic areas with the introduction of new very high-speed telecommunications. As a result, bottlenecks in access and processing are likely to fall back to the processors. This will make the computer companies the "fall guys" for performance problems in the future.

The demand for storage will accelerate at an exponential rate. Fortunately, technology changes in storage will enable gigabytes of information to be stored on devices that are no larger than a thumbnail.

Battery technology is also making enormous strides. Today's laptops would have been impossible to produce and use in the 1980s. Batteries then were too large. Some of us remember the batteries that were required for the first video cameras. Today, batteries for much more powerful systems are measured in millimeters and grams rather than feet and pounds.

Inexpensive processing and storage technology is very important for the use of access devices in underdeveloped parts of the world. Cheap devices with low-power demands, batteries that have a relatively long life in areas with no power and wireless/satellite telecommunications will enable the number of people accessing the Internet to surpass the billion-person boundary and reach multi-billion levels during the first quarter of the 21st century.

Table 1. Evolution of computers from the mainframe in the 1960s to pico devices in the next century

Levels of Computer	Location	Time Frame	Price ($000)	Numbers Installed
Mainframe	Data Center	1950s/60s	300	Tens/hundreds of thousands
Minicomputer	Department	1970s	30	Millions
Microcomputer/PC	Desk	1980s	3	Tens of millions
NC/IAD/PDA	Pocket	1990s	0.3	Hundreds of millions
Pico	Embedded	2000s	0.03	Billions

Global satellite technology will have a major impact on this. Low-orbit networks combined with low-cost access devices and inexpensive batteries will make Internet access easily available worldwide.

Whereas the personal computer used to be the only common device that had access to the Internet, there will be literally tens of different devices and eventually hundreds of devices that will access the Internet in the 21st Century. In the near future, these will include network computers (NCs), pagers, watches, PDAs (Personal Digital Assistants), IADs (Internet access devices), telephones, TVs and other displays, etc. In the not-too-distant future, access levels will shift from the NC level into the Pico level. At that time, there will be access from embedded systems. There will be complete phone integration and voice processing integration with Internet technologies.

Computing

There have been a series of waves of applications in the lifetime of the computer, as shown in the chart. There has been an oscillation between decentralized and centralized systems. These oscillations are driven by communications changes far more than by processing changes.

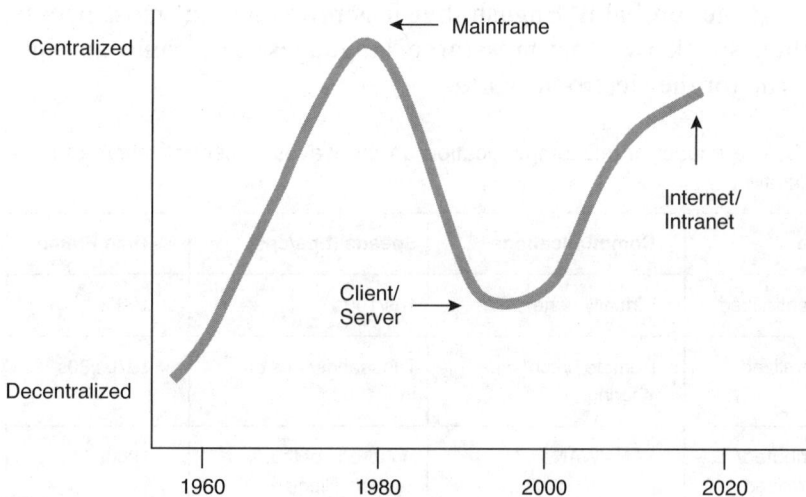

Figure 4. The waves of IT centralization and decentralization

The key difference between the second wave of centralization and the first is that in the second wave, all systems will not necessarily be located in one place or on one system, as they were in the first. In the first wave, the mainframe was jack-of-all-trades and the master of none. In the second wave of centralization like information will be centralized at one node in a network. There will be tens if not hundreds of such nodes devoted to processes and functions. They may be organized physically in any number of ways.

Perhaps all systems will be centralized into one node: this node may become a virtual node in the next wave and again be distributed through the network. This wave will probably occur between 2010 and 2020.

Many people keep hoping that we will simplify our technology infrastructure by reducing the number of operating systems, operating environments, languages, etc. In fact, the reverse is happening. We have more operating systems today than we've ever had. We have more information management and handling systems. The number of languages is increasing.

The thought that Java will become the sole language and the magic bullet that will solve all our problems is extremely naive. Java will be useful as a tool and a language for a set of tasks. It will also act as the umbrella for a variety of different languages. Humans do not all speak one language. Eventually, we may all have some knowledge of one language, probably English, but it is probable that most people will then speak two, perhaps three languages. The same analogy holds true for the electronic world.

Table 2. The impact of telecommunication on the waves of centralization and decentralization

Type	Communications	Speeds (bps/cps)	Time Frame
Decentralized	Virtually none	10s/100s	1960s
Centralized	Remote batch/ time sharing	Thousands/tens of thousands	1970s/80s
Distributed/ De-centralized	LANs/WANs	Hundreds of thou-sands/millions	1980s/90s
Centralized	Internet/Intranet	Tens of millions/ Hun-dreds of millions	2000s/10s

Telecommunications Transmission

Private funding of telecommunications capabilities is having a revolutionary effect on costs. As costs decline and available bandwidth increases dramatically, continuous series of thresholds are crossed. As each threshold is crossed, new Electronic Business opportunities will appear.

One result of the increase in both business and recreational travel is the spreading of ideas: When travelers see what is offered in one location, they become interested in having that same product or service available in their normal location. Governments can resist these demands for some time, but the more people travel, the more they are exposed to beneficial services, and the more their demands will increase. Eventually, any dam that is built will be breached. In many countries, the question is how long it will take to breach the telecommunications dams.

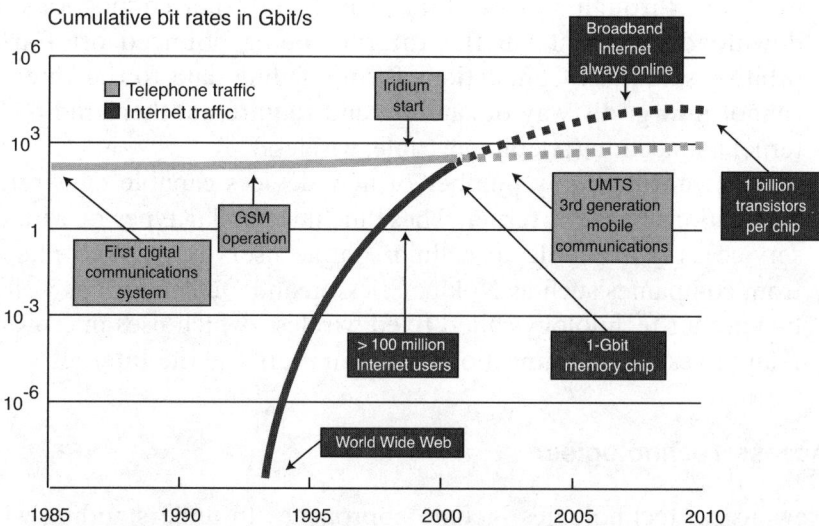

Figure 5. Milestones in the development of telecommunications

Over the next two years, bandwidth of half a megabyte per second to 1MB/sec will become available in dense metropolitan areas through various means. Information will be accessible at speeds 10 times to 100 times faster than today. Once people see what high-speed access (defined as anything dramatically higher than they have at home) can do, they will scream for it.

The pressure on telcos and governments to respond will be irresistible. Unlike earlier days, Internet users will be the mass of the population in many areas and in democracies they will vote to satisfy their needs.

However, telcos will have major opportunities in technology-based telecommunications services. Many of these services will be based on innovative software technology.

Connection Technologies for the Internet

Connection to the Internet is by:

- telephone lines (either over traditional dialup modems or by newer, faster modems that use a telephone wire-based technology called DSL, digital subscriber line)
- cable modems that connect devices such as computers and televisions over high-speed cable access lines to the Internet
- satellites, through services that provide Internet connections for downloading data from the Internet being bounced off Earth-orbiting satellites. Connections for uploading data to the Internet cannot now go by way of satellite and require another kind of Internet connection (telephone, cable, wireless).
- wireless, through any number of new devices capable of wireless connections to the Internet. These include similar types of wireless service as is available to cellular phone users (e.g., Web phones from companies such as Nokia, Ericsson and Qualcomm) as well as by another technology called fixed wireless, which uses microwave relays to establish connections between users and the Internet.

Access Technologies

New access technologies include approaches to understanding what the user is doing and wants to do. At the top of this list is ASR (automatic speech recognition). Its allows users to dictate documents and other written matters, as opposed to typing them on a keyboard. ASR systems also allow users to control devices, such as Internet browser software, by voice commands.

In the future, other interface technologies will enter the market to facilitate Internet access. These will include touch pads as pointing devices, gesture recognition, and even the kinds of eye movement tracking systems now used in "heads-up" systems in aircraft cockpits. Displays will include direct imaging to the retina.

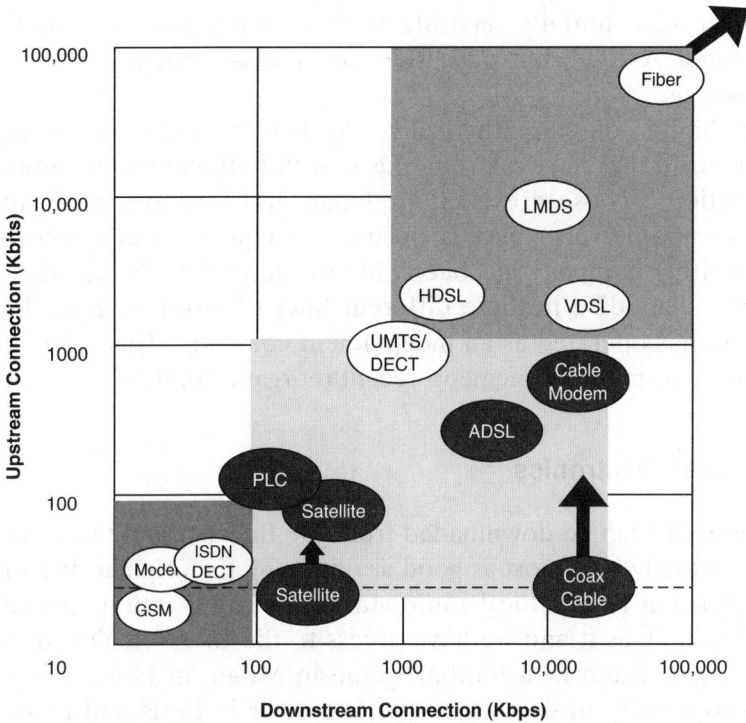

ADSL   Asynchronous Digital Subscriber Line   GSM   Global System for Mobile Communications
HDSL   High Bit Rate Digital Subscriber Line   UMTS   Universal Mobile Telecommunication Services
VDSL   Very High Bit Rate Digital Subscriber Line   PLC   Powerline Communications
DECT   Digital Enhanced Cordless Telecommunications

Figure 6. Transmission capacities of "last mile" technologies

Perhaps the greatest change in access will be in the number of sensor-based devices tied to the Internet. These devices will measure many different physical and biological properties. There will be sensor-based systems in homes, offices, factories, mines, ships and planes, all of which will be connected into the Internet. As we move into the 21st century, literally billions of devices will become connected into the network—and these are just the devices that are related to individuals, to people.

There will also be a great number of devices that are related to organizational activities. These will include switchers and routers, which are used for the construction, management and operation of networks. Until now, telecommunications switching networks and business networks have had separate development streams. These are coming together. Companies such as Siemens, Nortel and Ericcson were initially separate from organizations such as Cisco, 3Com and Ascend, but now there are cross-segment alliances and acquisitions.

We should note parenthetically that there are only two companies in the world that have a strong position in both computers and communications. Those are NEC in Japan and Siemens in Germany. Both companies also have a position in consumer electronics. To date, neither company has been able to successfully bridge the technologies embodied in these different lines of business. Each line of business has operated as an independent company. This is beginning to change, as noted by Siemens' recent reorganization.

Consumer Electronics

Digital radio can be downloaded from the Internet and the recording will eventually be almost as good as obtaining the original. We will be able to listen to our local radio station in any location around the world as long as it and we have access to the Internet. We can sit in Sydney and listen to a football game in Milan, in Hong Kong and listen to a rock music station in London or in Paris and record an opera being performed in Vienna. The quality of many of these recordings will not initially be high. But because the signals are digital and can be processed digitally, we will be able to have perfect reproductions available to us over the Internet. Copyright owners must deal with this issue.

Look at what will happen with technology applied to games, sports and events. By this, we mean not just games played by people interfacing directly with computers and networks, but also tennis, soccer, rugby and other sporting events, including horse racing, motor racing, etc. Technology will enable us to virtually participate in these activities. The miniaturization of cameras, sensors and microphones should enable us in the not-too-distant future to visualize and simulate a Wimbledon tennis match final.

Participation ability in sporting events roughly follows this sequence:

1. The ability to hear events and games wherever we are around the world over the Internet at our choice, not the choice of a broadcaster;
2. The ability to see and visualize what is happening;
3. The ability to "participate" as a player or contestant and choose the individual we wish to follow through the contest or the game;
4. The ability to not only choose to see, but also to sense what is happening.

Businesses built around these features may become some of the largest in the world in the 21st century. In the age of the Internet, audiences that have been measured in tens of thousands or hundreds of thousands might exceed millions, tens of millions and hundreds of millions. The impact on contests of all kinds will be dramatic.

Concerning electronic games, which are supposedly played by teenagers but often by adults, there already are game machines that are more powerful than any available PC. In a survey done several years ago on the interface of choice, the game machine interface was the most popular interface by far. It allows very rapid manipulation and control of the system. Game technology will be at the forefront of the development arena on the Internet.

Some people include content development as a technology. We consider it more as an application. There are two words that are used—"information" and "technology"—to form IT. Often, perhaps, too often, the emphasis is on the technology and not as much on the information. But in the near future, there will be dramatic changes in the constructs of information. Nicholas Negroponte at MIT talks about "atoms" of information. To use the physics analogy even further, we might consider that there will be information provided at all levels, molecular, atomic, subatomic, right down to the quark level. Assembly of this information into knowledge is what will be important.

Already there is more information on the Web than can be easily handled. In some areas the rate of addition is comparable to the rate at which one can access it.

Security

Electronic Business security will become one of the biggest industries of the 21st century. This will be quite consistent with historic precedence. In his book, *The History of Knowledge*, Charles van Doren points out that in the Dark Ages, up to the year 1000, as much as 75 percent of the income of people was spent on protection. They survived on what remained.

How much do we spend on security today? If all aspects of security were considered, from maintaining an air force to installing seatbelts, it would probably represent very significant proportions of our gross national products.

On the Internet encryption is a core security technology.

The United States has laws that treat encryption technologies as munitions and strictly control export, or even distribution by way of the Internet. The US software industry, and cryptography companies in particular, are lobbying the US government to relax these laws, citing the loss of sales of encryption technologies to overseas vendors not bound by US laws.

Without strong encryption, Internet commerce will be stifled by user concerns about the safety of their information, particularly relating to their financial transactions. Many vendors already provide strong encryption technology on their Web sites to address this concern.

Large investments will be required in the technologies of identification and authentication, protection, backup, and recovery.

We will apply the technologies of the Internet to our physical protection and other kinds of security. Already in urban centers, particularly in Britain, security cameras have become a very important part of crime control. In areas in the United Kingdom and the United States that are under constant video security screening, crime rates have dropped substantially. At issue is whether or not the crime is simply being pushed to another area, as happens when one pushes down on a water pillow. Indications, however, seem to be that this is not the case. If criminal activity can be made difficult, many more crimes will not take place. The detection and apprehension of criminals will dramatically reduce the incidence of crime. It is unlikely in the 21st century that we will move to a crime-free society, but we certainly should be able to enjoy a society that is much less crime-ridden.

Security devices of all kinds will be connected to the Internet. The first coordinated electronic attacks on national computer/information systems have occurred. We will see the equivalent in business. In fact, electronic security could be the Achilles heel of the Electronic Business industry. Again and again, companies tell us that they have run attack teams into their clients' systems that have broken security.

Again and again, they point this out to executives in these companies and are greeted with blank stares. Not until some of these executives are removed from office will proper attention be paid to security in this Electronic Business world. As always happens, there is a reaction after reported cases of sabotage, etc. Companies are often guilty of bolting the stable door after the horse has gone.

The IT industry, however, has done a fairly good job so far of protecting itself from security problems. Security breaches are always good for extensive press coverage so the small amount of press coverage devoted to security breaches today indicates that so far things are under control. This is despite the fact that the magnitude of business now being done over the Internet is greater than when the issue was first identified in the 1994–1995 time frame. At that time, many pundits predicted that Electronic Business would result in calamities, including bank failures, company failures, etc. These somewhat resembled the predictions of massive failure due to the Y2K phenomenon.

The Internet

This brings us to the Internet itself, which is not just about technology. The Internet was first proposed in 1964 and the first DARPA net node was established in 1969. It has been 30 years in development and has had major impact on our society as a whole. What is it and what are its characteristics? It is a network of networks. It is open, as DOS was in the 1980s. Unlike DOS, however, it is independent and universal. No single company owns it as DOS was owned by IBM and Microsoft.

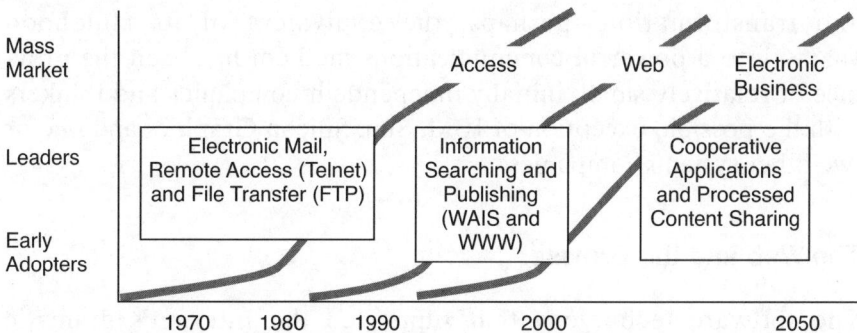

Figure 7. Three generations of Internet evolution

Why has the Internet been successful? Number one, it is distance- and geographically independent. It is successful in Australia, in the Arctic, Alaska, Austria, everywhere where there is any degree of freedom. It is owned neither by any one organization, nor by any one country. It is not, and from very early days has never been, an American construct.

Number of Internet users
(in millions)

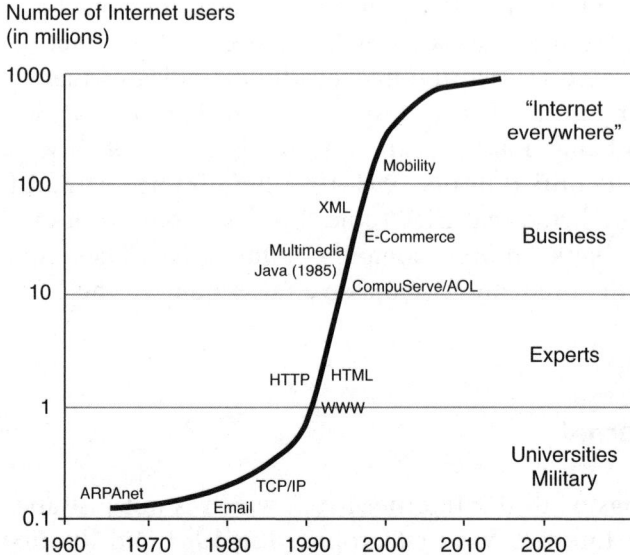

Figure 8. Key technologies driving the Internet

It is an environment that is driven by demand, not supply. None of the innovations and initial developments that made the Internet at- tractive was carried out by major companies or organizations. It is true that the Internet was conceived, developed and born of US gov- ernment initiatives. Once it was born, however, its nurturing, growth and transformation—perhaps, the equivalent of its childhood years—into a powerful communications medium has been the prov- ince of relatively small, initially independent companies and thinkers with the possible exception of IBM, Sun, Silicon Graphics and one or two other systems companies.

The Web and the Browser

The software technology that supported the Internet's dramatic growth is the universal access method or HTML (HTML = hypertext markup language). This language is easy to learn and can be implemented on almost any computer. Tim Berners-Lee first put this technology to successful use while working at the CERN physics facility in Switzerland. Berners-Lee thought it would be marvelous if all physicists could readily communicate, despite using different computer systems and software, and could link their communications and documents to each other. Thus was born the World Wide Web (WWW or Web), that eventually spread beyond physics to embrace the entire world.

Hypertext transfer protocol instructs the Internet in how to locate and convey the information to any place on the Internet. Hypertext markup language provides a code to use to instruct any HTML-compliant browser in how to arrange and display the information on a Web site onto a computer or other display screen.

The second software technology component that made universal access available is the browser. This is the piece of software that sits on client computers and interfaces with the software to access the network. Built upon the work of the World Wide Web by graduate students at the University of Illinois and the US National Center for Supercomputer Applications, Mosaic, the original browser was meant to let Web users view multimedia content. What made this new technology successful was its way of accessing and distributing content to varied and incompatible computing and telecommunications systems.

Thus, in the creation of content, it is unnecessary for the content provider to know whether the browser at the client end is from any specific manufacturer. This tremendous freedom enables a content developer to know that any client can access the information provided they used standard markup languages. It is the primary criterion for the success of the Internet and the World Wide Web. Unfortunately, there are signs that individual vendors wish to put their proprietary stamps on browsers, thus producing an unwelcome level of incompatibility.

Everybody's a Publisher

The freedom to publish is another reason for the Internet's success. Given the above standards of information preparation, distribution and access on the Internet, it became easy for almost anyone to be an Internet publisher. Unfortunately, this meant that a lot of information on the Internet is "low value." But value, like beauty, lies in the eye of the beholder. It becomes very difficult to judge the value of information. What may be totally useless to 99 percent of the Internet population may be very valuable to the remaining 1 percent. That 1 percent represents 2 million people. A market or user community of 2 million people is a very significant one by almost any standard.

The Web and the Internet also means that people and organizations with "high-value" information but no access to standard media can disseminate their knowledge. As a result, electronic newsletters of all kinds and chat or affinity groups have become common on the Web.

The Internet is a truly open environment. It is quite simply the world's biggest and most successful open system. By definition, an open system is one in which any component can be replaced by a component from another vendor without interfering with the performance of the system. All Internet standards are public. Its openness is the biggest reason for the Internet's growth.

In this context, the Web is really an application enabler, not the Internet. The Internet is the delivery channel, thanks in great measure to IP, or Internet protocol (often referred to as TCP/IP), which is the standard for communications carried over the Internet.

The Web did for the Internet what spreadsheets and word processors did for PCs: It made the Internet usable. The Web provided a reason for the Internet's use above and beyond that of simply communicating through e-mail systems. There's no doubt that for the simple access and communications capabilities alone, the Internet would be reasonably successful. But it is the Web that has driven its growth and started to change us into a truly information-based society.

Now, Berners-Lee and others are championing what is expected to be the Internet and the Web's next important tool: XML, or extensible markup language. XML will be used to supply information about information on the Internet: it will help in accessing information that today is hidden away in databases.

It allows Web page designers and creators the ability to include new data types in information on a Web site, for example, material from databases. Though not widely used at present, it will grow substantially over the next three to five years, as Electronic Business becomes a far more dominant net activity. XML has the potential to provide prospective customers with searchable databases full of information about product features, pricing, availability and other important information.

Chat technology is a runaway hit on the Internet. Chat involves delivering software via the Internet that lets users "chat" with one other in real time. The chatting is done at the moment via text typed on keyboards, but the real-time element of chat is what makes it so compelling for users.

Business users have embraced chat as a means of achieving collaborative work goals. Since much chat software allows users to see the nicknames or e-mail addresses of those who are live online in real time, teams can assemble online and use the Internet to have a text-based discussion or meeting. With the advent of more advanced Internet user interface technologies (see below), voice and video may well replace text as the main media of communications in Internet chat sessions.

Internet telephony, also called VOI (voice over Internet), takes the traditional analog signals that are sent over copper phone wires, converts them to digital information and sends them as packets over the Internet. These packets are reassembled at their destination, converted back to analog signals and the phone conversation takes place, often with surprisingly little delay. With the advent of higher bandwidth Internet services, VOI may well replace much traditional analog telephony traffic.

The Internet as an Enabler of Technology

The Internet is not only dependent on technology; it is an enabler of the application of technology to business and other uses. In the external Electronic Business world, the Internet enables the use of all kinds of software technology, those related to searching and information access having become particularly important over the last few years. Literally, thousands of software companies have and are being formed to develop software technologies to apply the Internet for both internal and external purposes.

The following sections discuss how the Internet and Internet technology are enabling changes in technology use for internal as well as external systems.

Migration from Legacy Client/Server to Internet Technology

Computers began life in the business world as special-purpose, stand-alone tools, operated and maintained by a select priesthood. One waited on queue to merely submit a request for work on the big monolithic machines in the air-conditioned rooms or "glass houses."

Then came the eras of minicomputers and microcomputers, leading to a distribution of computing power throughout the organization. Client/Server systems emerged as a means of using and managing distributed systems of minicomputers or servers, local area networks (LANs), and desktop systems. They enhanced and in some cases replaced centralized systems. Most organizations now have a legacy environment of mainframe and client/server systems in place as they prepare for the Internet.

"Intranet" is the application of Internet technology to internal systems on what is essentially a VPN (virtual private network). Organizations use servers, the IP protocol and browsers to interface internal staff with each other and applications systems in central, divisional and departmental locations. They can also be extended to partners and associated staff in a networked organization.

Approximately 80% of major US organizations either have implemented or are implementing Intranets and the growth in their use is accelerating. Intranets will become the information and knowledge hubs of organizations.

At the moment, users are focused on interfacing client/server technology and Internet-based technology. They will switch to native mode Internet applications early in the 2000s.

Internet systems will be different from mainframe and client/server systems in a number of important ways. Earlier systems are two- or at the most three-tiered operations, while newer Internet-based systems will have any number of tiers. Internet applications also introduce a middle tier, which allows users to use any type or combination of standard database solution inside or outside their organization. The mainframe and client/server approaches were based on proprietary protocols, while the Internet/Intranet approach is based on open systems, especially Internet standards like TCP/IP, HTTP and HTML. These kinds of standards allow the Internet/Intranets to offer almost complete computer, communication and accessory platform independence.

Stand-Alone **Client/Server** **Internet/Intranet**

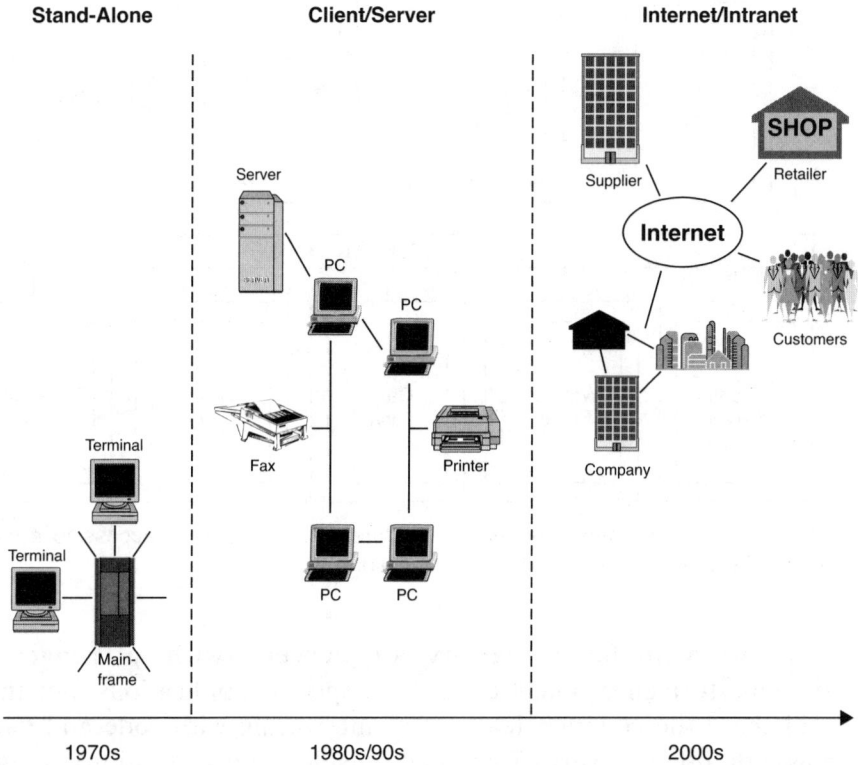

Figure 9. The development of networks and the changing technology focus of IT platforms

The Internet will offer many benefits to users of next-generation, native-mode Internet applications, including application flexibility, scalability, re-configurability, open application interaction, continuous application enhancement, applications extension beyond the boundaries of the enterprise, and improvements in software maintenance and re-usability.

Next-generation applications will use networked servers upon which reside business applications or objects to be accessed by "clients" (PCs, PDAs, computers, other devices) across the Internet. Such "Application Servers" can be anywhere; they may be centralized or distributed by geography or by business unit.

File Server	Application Server	Database Server	Web Server

Internet (TCP/IP Network)

Notification Server	Watcher Server	Game Server	Video Server	Transaction Server

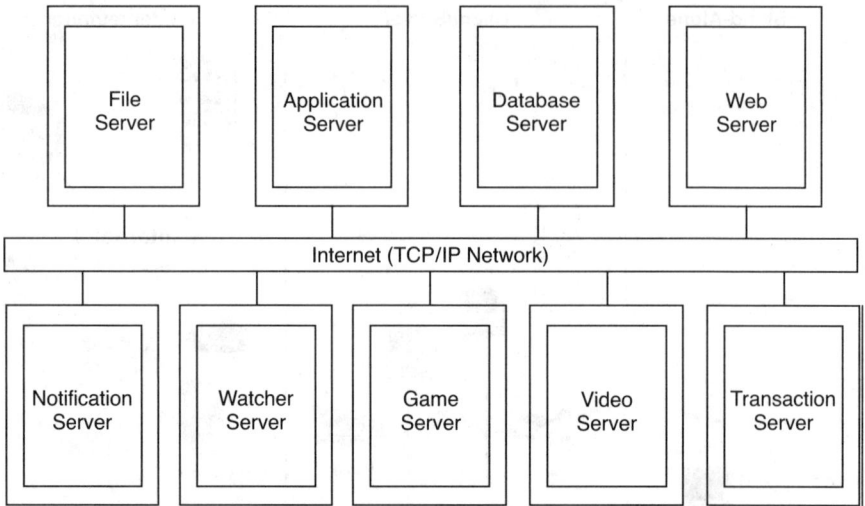

Figure 10. Internet server types—In the Internet, clients have access to a wide range of servers, which deliver defined functions

There will be a variety of servers with a range of price/performance points tailored to specific applications, but they will share the common features of interfacing with code and databases on other servers (inside and outside the organization) and delivering application services to client devices. These servers will include:

- **File Server** – This is a directory server such as Novell's NDS that will be mapped onto TCP/IP for directory services over the Internet. There will no longer be a necessity to "drag and drop" files in a browser or other interface.
- **Database Server** – This is a backend server that will process database activities; they will be object-oriented. There will be different servers for different types of objects. Companies such as IBM, Oracle, Informix, etc. provide the software to operate these servers.
- **Web Server** – These servers are already common and come from many different suppliers. A key capability is security although some systems may put a security server in front of a Web server to manage the firewall activities. Another key capability is scalability, the ability to handle very rapid changes in access volume. Netscape, Microsoft, Open Market, O'Reilly & Associates, Process, etc. are suppliers of these systems that will service HTML documents and objects.

- **Sensor Server** - Sensor servers handle Web clients' measurements such as temperature, pressure, inventory levels, energy usage, etc. over the Internet. The clients may be very small devices that provide data via wire or wireless communications and have very small Operating Systems to run them. Similar to Web servers, sensor servers will answer GET requests from a Web client and will return a command, or they may poll clients, or they may monitor continuous readings and respond to out-of-range conditions. They are similar to and can be process control systems.
- **Games Server** – These servers deal in high interactivity and visualization. They may also have to adjust to large variations in group activity as multi-player games develop. Server software will come from Sony, Sun, Sega, Nintendo, Electronic Arts and others; often these servers will be collaborative involving companies such as Silicon Graphics, IBM and Siemens in providing the technology to support these leading edge applications.
- **Transaction Server** –The transaction server will divorce transaction management software from the Web server. It may also be a data base server or may sit in between the Web server and the database server, particularly if the database server is also supporting non-Internet activities. Open Market's OM-Transact product is an example.
- **Audio/Video Server** – These servers primarily support entertainment activities. They are not highly interactive but must support heavy one-way communications requirements.
- **Mail/Collaboration Server** - These servers handle e-mail messages and other forms of interpersonal communications, including collaborative activities such as chat. As e-mail proliferates it will demand separate attention. As video is added to the mix, specialized servers will handle videomail and videoconferencing.
- **Other Servers** – Other kinds of servers will be attached for special needs.

Because applications are independent of the target client device, an application can be dynamically downloaded, without modification, to a broad range of client devices including "smart" telephones, point-of-sale devices, PCs, workstations, Internet appliances, set-top boxes and more. About one fifth of the client devices by value (excluding PCs) purchased by organizations will be Internet Access Devices (IADs) or network computers by the year 2003; they will be far higher in terms of numbers.

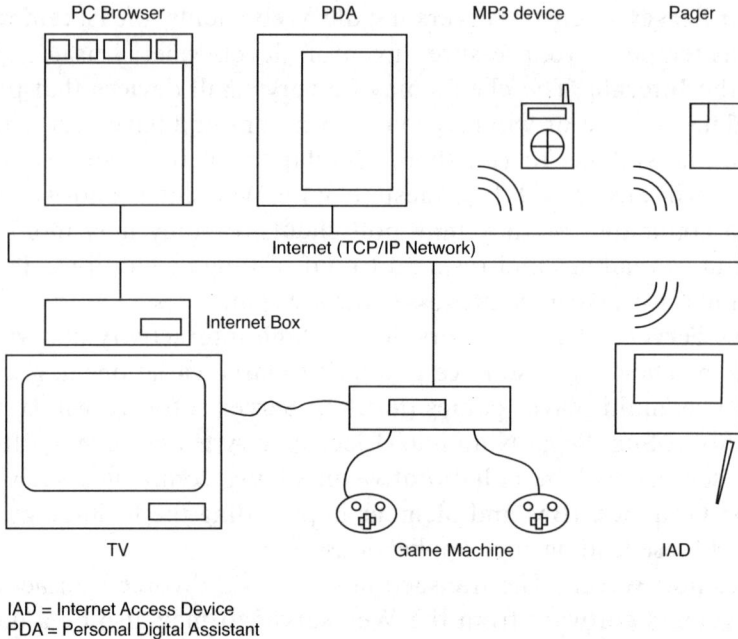

Figure 11. Internet client types

Internet-generation enterprise software will have to be implemented using object technology because current procedural methods and tools cannot support the complexity of the operating system and network management needed to control the infrastructure.

The Internet and object browsers will accelerate the movement toward the use of objects to create network-aware applications that extend beyond the boundaries of the corporation.

The impact of these changes will be as great or greater than the impact of the client/server revolution at the end of the 1980s. As a result, the IT infrastructures of most organizations will have to be changed. This will be anathema to many IT directors who have been able to move from legacy mainframe solutions to C/S only in the last five years or so.

Each major change in architecture has led to an acceleration of the switch from in-house applications to outsourcing and other contracting methods. The move from C/S to Internet will be the final factor for many organizations. Variability in telecommunications resource costs, features and availability will make Electronic Business applications as dependent on telecommunications capabilities as on computer and software capabilities. Volatility of applications will be much greater in Internet Electronic Business applications than in either of the two previous structures. It is difficult to imagine applications that can remain "frozen" for 5 years let alone the 20 or more years that some have remained in the mainframe world.

Table 3. Phases of Intranet use

Phase	Description	Examples
One	Static information distribution, mostly administrative	Company policy documents, staff and telephone directories, and visitor registers
Two	Business unit and departmental information sharing	Product plans, financial data, customer service records and sales contacts
Three	Group collaboration	Project management, groupware and desktop conferencing
Four	Integration of existing systems and application with Internet	Web-enabled data warehouse, Web frontend to legacy databases, product design and live customer service querying
Five	Replacement of legacy systems with Intranet equivalents	All current applications

This will create an opportunity worth hundreds of billions of dollars worldwide for new platforms over the next 10 years. This change, however, might reduce the potential for very large platform projects because the Internet by definition allows the interface of dissimilar applications, server structures and networks. Therefore, it will be possible to break up the monolithic networks that have traditionally been the platform for numerous applications and replace them with numerous tuned networks based on the Internet and the ubiquitous browser interface.

We will not return to the days of the mainframe. Centralization will be built around function and process; servers, which may be very large computers indeed, will be dedicated and specialized as we indicated above.

Table 3 shows the phases for Intranet use. Most large organizations are at about the Phase-Two level, with some activities at Phase-Four levels. Many of these organizations will reach the Phase Four level by 2000. They will take another five to 10 years to shift to full Internet-enabled internal applications. This is similar to the addition of C/S systems to mainframe systems.

Most projects at Phase-One and -Two levels are at the hundreds of thousands of dollars level in cost. By Phases Three and Four, projects are reaching the multimillion-dollar level and Phase Five projects are at the tens of millions of dollars level—for large companies.

Software

Software technology is of utmost importance. Computers are often simply software-on-silicon, or SOS. Semiconductor companies such as LSI Logic will tell you that they actually employ more software engineers than electrical or electronic engineers.

Application Development

Development and management tools for Internet applications are becoming available. The use of these tools and structures will reduce the complexity of applications. Rather than have to develop systems where the complexity is in designing module and systems interaction, designers will be able to optimize the structure of each application independently. Interfaces will then be commonly and easily available to other applications and structures.

Most Intranet projects so far have been less than $1 million in size and often carried out by new, small organizations. This is changing as the power and economy of the Internet technology grows. The average size of these projects will increase and eventually the core, mission-critical systems will become Intranet-based, rather than today's situation in which Intranet applications are usually just access systems.

Internet applications are following the same path but expenditures are now overtaking Intranet expenditures as companies focus on Electronic Business applications and not just presence on the Web.

Overall, the trend will be toward many smaller projects rather than fewer larger ones. Of course there will always be very large applications projects in very large organizations; but, in the future, projects will be more often for pieces of a process rather than a whole process. Users will be able to deal with process problems one piece at a time. This should make development faster, easier, cheaper, more flexible and more satisfying

Development Tools

One of the most important areas of software is that of development technologies themselves. We want to build applications from objects that represent building blocks, and have an assembly process as opposed to a custom-manufacturing process.

The tools as yet have not been that effective, however. They have been proprietary, and users have been loath to put all their eggs in one basket by selecting one or another methodology.

Developments like Java (described below) and the Linux operating environment hold out hope of development standardization that will greatly benefit the industry. We foresee a time of user-defined software. Individual users will specify the type of activities they wish to perform and essentially have software constructed in an interpretative basis that meets their demands.

The Impact of Java

Java will play a pivotal role in defining the shape of Internet computing through 2003. Figure 14 shows the major elements of a possible Java-oriented Internet/Intranet environment. Thickly outlined elements are those most directly impacted by future changes. An Internet environment will connect to other environments outside the organization in an unstructured manner, whereas the Intranet environments will have controlled internal connections.

Note that this definition does not exclude connection of an Intranet to external partners, whether contractors, other suppliers or customers. The difference between an Internet and Intranet structure is largely one of predetermination of connections: Internet is ad hoc; Intranet is predetermined. There are many gray areas and an Intranet system can easily become an Internet system in many cases.

Whether Java remains as a strict development standard or becomes a series of compatible "standards" could seriously affect the nature but not the size of the IT industry.

```
                              ┌─────────────────┐
                              │   Java Objects  │
                              └─────────────────┘

  ┌──────────────────┐
  │   PC/Workstation │
  └──────────────────┘
                                                    ┌──────────────────┐
  ┌──────────────────┐                              │    LAN Server    │
  │ Network Computer │                              │    e.g. mix      │
  └──────────────────┘                              └──────────────────┘

  ┌──────────────────┐      ┌──────────────────┐    ┌──────────────────┐
  │  Java Terminal   │      │   Java Virtual   │    │    Intranet      │
  └──────────────────┘      │     Machine      │    │     Server       │
                            └──────────────────┘    └──────────────────┘
  ┌──────────────────┐
  │   Java Devices   │
  └──────────────────┘

  ┌─────────────────────────────────────────────────────────┐
  │              Security Infrastructure                    │
  └─────────────────────────────────────────────────────────┘
  ┌─────────────────────────────────────────────┐       ┌──────────┐
  │      Internet/Intranet Infrastructure       │───────│ Firewall │
  └─────────────────────────────────────────────┘       └──────────┘
```

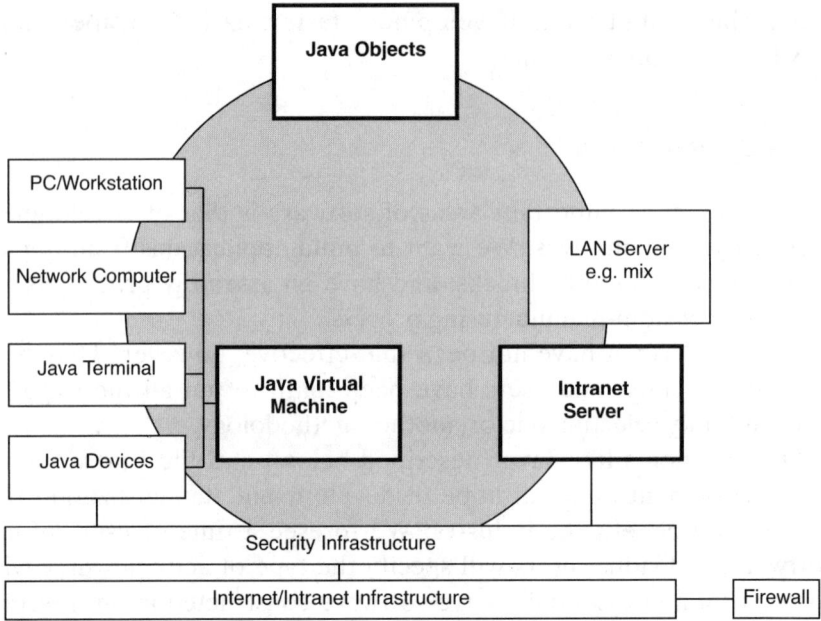

Figure 12. Major Elements of a Java-oriented Intranet environment

If it is a strict standard, the industry might arguably be reduced because of the easy availability of standard Java-based applets and applications. In this case, however, the very ease of use will convert to more opportunities. Thus, individual activities and projects may be smaller on average, but there will be many more of them.

The speed of change of major systems will accelerate, providing more opportunities. If the building blocks are easy to assemble, companies will find more ways of assembling them.

We expect that almost all client and server platforms (including specialized Internet appliances and embedded devices) will be delivered with Java capability. This will simplify some of the software development activities. Most important, applications will be decoupled from the underlying operating system and hardware platform.

If Java behaves as Unix today and is promoted as an "open" environment, but in reality becomes a set of fragmented, proprietary implementations of an "open" system, then the market will resemble today's market even in the new Electronic Business areas. Fragmentation will occur throughout the emerging Internet/Intranet environment, with vendor- and platform-specific implementations caused by suppliers' contrary desires to standardize, yet retain or obtain an advantageous market position.

Development skills will shift to the Java model, but delineation by vendor's architecture, as is the case today, will continue. Also, changing to a new language is always more difficult than it seems. It is necessary to have a complete set of tools, processes and procedures in place before major applications development is undertaken on a broad scale. Organizations simply cannot risk failures in the Electronic Business world.

Java might not achieve widespread success due to an escalation of problems, including slow performance, irresolvable security issues, and inflexibility caused by lack of access to local client resources. In this case, code compilation will remain native (specific to the hardware and operating system), and applications will remain tied to the underlying platform. Systems Integration will then be even more important and projects will continue to be large and relatively slow to implement.

Coding the Process

The phenomenal growth of enterprise applications solutions companies, particularly those in ERP such as SAP, Baan, Oracle, JD Edwards, etc., is another aspect of the software environment. One must pose the question why these companies, particularly SAP, were so successful. In our view, it was SAP's recognition—which may not have been explicit at the time—that it was not really involved in coding an application for the computer. What it was really involved in was coding of the knowledge of the particular process that the computer was supporting. Initially, SAP would not include modifications to its software, nor allow them, unless a number of potential buyers supported one buyer's suggestions. If this support occurred, then SAP would code the change into its software. This meant that every process was tested not simply in one company, but in several before it was embodied in the code. This enabled the knowledge process in the code to become a best-of-breed process activity, a body of knowledge about business processes. Application software changed in the 1980s and '90s from being the coding of instructions for computers to being the coding of processes.

Recently the ERP and other application vendors have been hit by a "double whammy" in the market. Firstly we have entered the Y2K systems recession in mid-1999 where organizations have frozen their platforms until the New Year has passed and/or they are sure they can handle it. The second impact is from the reorientation of systems priorities from internal productivity to external business generation. The latter move has benefited companies such as Siebel Systems but has hurt the ERP vendors. Nevertheless the principle of coding the process is sound.

Application developers are now inventing processes. Consider the terminology used to describe people on the cutting edge of the Electronic Business world. We do not talk about programmers and analysts; we talk about software engineers, architects, designers, Webmasters, etc. These are all creative titles. Programmers and analysts were non-creative people in one sense, while being particularly creative in another. They were brilliant at collecting information on what was done, analyzing and improving it, and writing computer instructions. Today's developers are focused not on what was done, but what can be done.

The Future

It is almost trite to say that we will have supercomputers on a laptop. What is perhaps more interesting is that we will have supercomputer processing power in a pair of eyeglasses. Much of the personal electronics that we use will be wearable, like watches or jewelry. While we do not yet expect to see the equivalent of the *Star Wars* personal transportation system on our wrist, we will certainly see its equivalent in terms of our ability to communicate with the Internet, and through it to other businesses and individuals.

Very high-speed communications access to the home via fiber, satellite and enhanced wire technologies is very predictable. Storage technologies that enable us to keep our complete set of personal ID and records on a storage medium as small as a coin are envisioned, as is a ubiquitous availability of computer connection facilities. Laptops and portable devices may simply disappear, because everywhere we go, we will have secure access capability. Our hotel rooms, our offices, our cars, our airplane seats will all have the equivalent of a laptop or IAD with display capability. Already some hotels, such as the Inter-Continental, provide Internet access capabilities as part of their executive rooms. If a person can instantly interface their wearable system and device to the local system and is not charged exorbitant prices for doing so, then he or she will find it burdensome to carry around a two- to five-pound piece of luggage that is expensive, fragile and easily lost.

Technology advances in the software environment will be truly amazing. As well as software for agents, which has been well explored in the media, we will have software for content video; thus, we will attach video to messages, products and services of all kinds. Our eyes have the highest bandwidth of information input to our brains of any of our senses. We use a very small portion of them today when we look at screens of static data. 3-D visualization will become extremely important over the next five years. This will be accompanied by various changes in the input processes. Voice and translation capability will be important, but there will also be motion indication and gesture recognition. There are experiments with brain wave control. People can already sit before a system and adjust the position of a cursor on the screen merely by thinking, using superconductor-enabled skullcaps.

Almost all IT development will be focused in and around the Internet and the Web. This will be a worldwide phenomenon. It will be the new Holy Grail for technology developers.

4 Electronic Business Barriers

The major barriers facing entrants into Electronic Business are presented in this chapter. The decisions accompanying entry into the Electronic Business domain must take account of significant challenges and difficulties. Many of the criteria for making the right decisions are clouded by the sheer pace of change and speed of developments related to the Internet and other aspects of Electronic Business and technology.

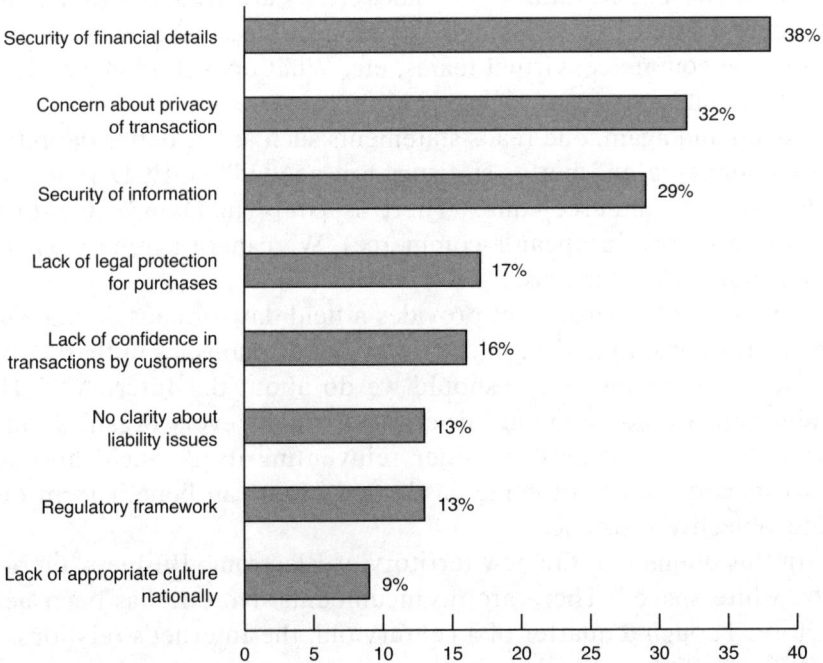

Barrier	Percentage
Security of financial details	38%
Concern about privacy of transaction	32%
Security of information	29%
Lack of legal protection for purchases	17%
Lack of confidence in transactions by customers	16%
No clarity about liability issues	13%
Regulatory framework	13%
Lack of appropriate culture nationally	9%

Figure 1. Barriers to Electronic Business (Andersen Consulting)

This chapter reviews the knowledge and resources required to overcome these barriers to Electronic Business. The issues presented focus first on business areas, both organizational and operational, that will affect the decisions that executives must make in implementing Electronic Business. After that we present a review of more specific topics such as financial issues, technology concerns, language barriers, security and interoperability, as well as the usefulness and achievement of technical and Internet standards.

Fundamental Issues

Knowing What to Do

Lack of knowledge and experience form the biggest barrier to conducting Electronic Business. In major companies in the United States and elsewhere, senior executives are scratching their heads and saying "Holy s---, what do we do about this?" and "When do I do it?"

The world has changed. Suddenly, everything is happening in different ways. The headlines in business news are all about things that scarcely existed five years ago, let alone a decade ago: Web sites, portals, e-commerce, virtual teams, etc. What does it all mean? How to sort it out?

Again and again, one reads statements such as "A better definition of e-commerce and better statistics are needed" (OECD report on Electronic Commerce) and "There is [*sic*] <u>No</u> Data," (Goldman Sachs paper on European e-commerce). We cannot agree on what it is, let alone what it means.

This kind of environment provides a field day for consultants. One major US consulting firm said, "Every major publisher in the US has come to us asking, 'What should we do about the Internet?'" The publishers are not alone in this quest. Virtually every organization is being forced to at least consider reinventing itself. Such thorough introspection does not come easily and often can benefit from outside, objective assistance.

In this domain, in the new territory of Electronic Business, there is all "white space." There are no incumbents. No one has been here before. Though a quarter of a century old, the Internet's relationship with business began only a few years ago. As a result, what can you trust? What should you regard with suspicion? If everyone is new to this environment, if all are pioneers, including consultants, whom can you trust? And the decisions to be made are "bet your company" decisions.

There is a battle in the temporal position: First movers in any space rarely are dominant after a few years, much less in one so rapidly paced as the Internet. Yet if they get too far ahead, it will be hard to catch them. Where will Yahoo!, Amazon.com, MSN and Lycos be in 2002? Are they too early for the wave, or are they riding the crest of it? These companies, with the exception of Amazon, are "arms merchants" to the Electronic Business industry, rather than Electronic Business companies per se. Who will be the true Electronic Business companies in five and 10 years? Where will the early Electronic Business companies such as Schwab and Amazon be?

The other side of this equation is knowing when the Electronic Business wave will pass. Like a surfer, if you miss the wave, you may never catch up. Internet time is a cruel and unyielding new measuring stick. The amount of time to assess, plan and act has been shortened by all of the things that Electronic Business makes possible. There is less time for sober assessment than in the past, yet serious action and perhaps more so inaction in this new era is accompanied by much greater risk. By embracing the Electronic Business environment, a company can suddenly be thrust into a marketplace in which it has no experience and little knowledge base.

Among the thorniest problems and the stickiest decisions in the Electronic Business domain is knowing how much to invest and knowing what kind of commitment to make. The two are intricately intertwined. The level of investment made to embrace Electronic Business solutions is directly linked to the size of the enterprise and the nature of the solutions being pursued. The nature of commitment is more directly related to confidence factors.

In the financial industry, two brokerage companies pursued different paths: Schwab committed; Merrill Lynch didn't. It hesitated for many months, with no commitment, and even made numerous public relations pronouncements saying that stock trading on the Internet was not *worthy*.

eSchwab prospered as the vehicle through which Schwab embraced the Internet. It offered customers any number of choices, including trading with a live broker at a branch office, trading with the broker by telephone contact and trading online, over the Internet. eSchwab went even further and committed to make financial and equities data and information available to its customers through any number of modes, but especially over the telephone, through new telephony data services and online.

Merrill Lynch capitulated late in the spring of 1999, admitting that a significant number of its customers were interested in using Internet-based trading processes and that it was not servicing those customers and that market. Merrill was hounded into a philosophical reversal of its former dismissive stance by eSchwab and upstarts such as Etrade, AmeriTrade and the other new online brokers. It announced that it would begin to offer Internet trading at the end of 1999, with trades priced, for Merrill, at bargain commission rates that would start at $29.95 per trade.

So, not every firm that has embraced the Internet in its business model has achieved success. Just sticking the Internet onto existing business processes will not be a successful business model; it will not be enough. A full commitment requires research, investment, timing, recruitment of new talents, training and the development of a suitable business model reflecting the new technologies and processes that are being embraced.

Crossing the Chasm

This chart presents the chasm inherent in embracing Electronic Business. The case described above in equities trading is an example of different approaches to cross it.

Merrill Lynch brokerage is on the left side of the chart, using e-mail and Web sites to amplify its regular way of doing business. eSchwab is on the right, using the tools of Electronic business in earnest.

The key here is how and when to move across the chasm from restrained use of specific tools to a fully integrated Electronic Business.

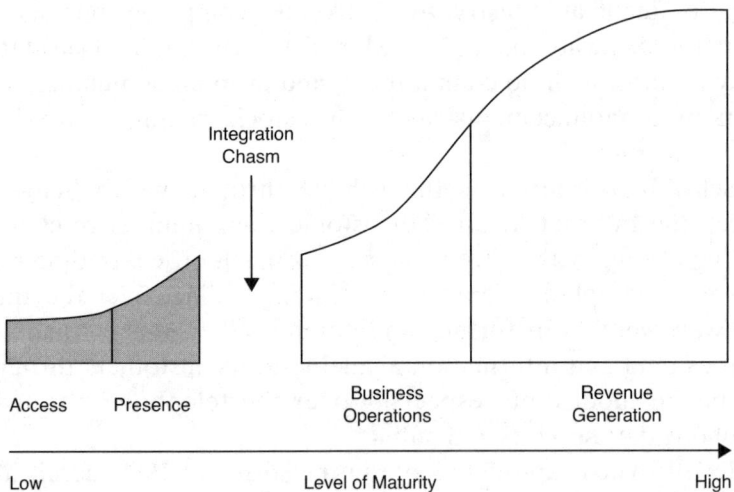

Figure 2. Electronic Business adoption cycle

This is the huge barrier. Access (messaging, e-mail, etc) is easy. Putting up electronic billboards and online versions of paper brochures is a start. It requires minimal effort, expenditure and commitment. It might even boost recognition and reach new, online markets. It does not create an Electronic Business. This requires presence.

Presence is harder. It involves Web hosting, Web marketing, Internet support, elementary e-commerce using minimal systems, security and privacy protection implementation, etc. It includes determining the new ways that customers and prospects (whether they are individuals, businesses or both) will be served by your Electronic Business processes. Essentially, this is analogous to building a pilot plant.

The chasm is moving from these secondary, experimental pilots to implementation. Putting all your inventory online rather than just some selected items and contacting all your customers to offer them this alternative channel. This transfers Electronic Business principles into working business operations and, eventually, full-scale revenue (and profit) generation.

This comprehensive changeover involves all of the planning and work necessary to put into place the infrastructures, processes and applications that make realization of the new Electronic Business opportunities possible. Most important, it means providing new kinds of customer services that support the Electronic Business environment. All of this is expensive and extensive in its reach. For example, even keeping Web sites current is an enormous task, especially when you consider that performing a complete rewrite can take up to six months. The chasm is crossed when this approach is adopted in the enterprise as a whole.

Integration and full-scale operations in Electronic Business are very large problems, with many complicated issues. "Commitment" becomes the key word because of all the resources needed and the management required for those resources. This could be where and when you bet the company.

To date, companies have looked to IT to find solutions to these problems and implement Electronic Business. As measured by IT departments and consultants, the impacts of implementing an Electronic Business are largely technological.

But in the last six months, company marketing departments have taken over responsibility for Web. They have recognized the importance of the Web site in positioning their company and in communicating to the world. Unfortunately, many of these companies' marketing departments do not have the combination of technological knowledge and marketing that this requires.

Getting the Right People

Labor is another large barrier. Prices of Web literate designers and implementers are among the highest in the industry: $1,000 to $2,000 per day—if you can get them.

The value of such people is enormous. One senior industry executive recently put a value of $2.5 million per head on a group of top-level Web/Internet technology specialists out of one of the leading US universities—a 20-person company with no revenues and a value of $50 million!

The labor shortage is real and will continue to grow. No educational infrastructure is in place to educate and train for the delivery of these new services. Many learn on the job or by purchasing, studying and implementing the instructions in how-to books. Once on-the-job learning is completed, greener pastures with larger salaries and other compensation items beckon. One of the things to be especially alert to is keeping employees who have acquired these new Electronic Business skills satisfied and devoted to the organization.

There is a war for talent. The most talented, well-informed and valuable employees, the ones who have learned these new tools and techniques, are the ones who are most likely to be wooed away from their existing employment. Recruiters and headhunters specializing in Electronic Business are increasing in number. Many use the most aggressive techniques to discover who is responsible for creating the most appealing Web sites. Those individuals are then approached and offered more money, more options, even equity stakes and other forms of inducement to leave for "better situations."

Table 1. Ten new e-commerce jobs (OECD Report, 1998)

Job Title	Tasks
Entrepreneurial consultant	Analyzes the overall business case for a project and turns around struggling enterprises. Part merchant banker, part visionary, part technocrat, you force your clients to rethink their place in the world, then re-engineer their business.
Application developer	Creates new software programs or online business tools. New businesses require people to create (develop) the

	structures (applications) to help them succeed. This may be a new Web site selling technique or a way to share company information among employees.
Fulfillment specialist	Gets the product to the consumer.
Consumer behavior consultant	Analyzes why people buy things. The AC Nielsens of e-commerce. With so many people using the Web in so many different ways, it is necessary to have adaptive, meaningful measures of success. Someone who can evaluate consumer behavior can help an enterprise better target its audience.
Broker	Finds new business opportunities and staff; a recruiter. As an employment broker, you can expect to get 20 percent of the talent's first-year salary in commission. In return, you will find the people from the other nine categories listed here, many of whom will not have direct IT training, but complementary skills that can translate to e-commerce.
Network security specialist	Makes sure that computer systems are safe from prying eyes.
E-commerce business analyst	A bean counter; a number cruncher.
Internet architect	Puts it on the Web. The people who design the site and conceive concepts. A Webmaster controls the team that puts the pages online, like an editor for a newspaper or magazine.
Product manager	Makes sure it stays on the Web. The environment is constantly evolving and e-commerce products need to be kept on track. The day-to-day programming of the Web needs a timekeeper.
Core programmer	Takes care of day-to-day computer programming tasks.

Who Do You Do It With?

Lots of helpers are available. "Specialists," even in these new special-ties, abound. Many large companies such as Hewlett-Packard, IBM and Siemens Business Services offer varieties of products and services for Electronic Business implementation. Major consulting firms also offer services that range from e-commerce turnkey solutions to hand-holding in the transition from enterprise to Internet systems.

What is the optimal balance between internal and external resources? External organizations generally still have little practical
experience in Electronic Business. Internal organizations have none
and may well be resistant. In every organization today, however,
there are people who are charged up by what is happening and are
eager to find ways to apply Electronic Business to the organization.
Find these people and establish entrepreneurial teams entrusted with
carrying out the organization's Electronic Business planning and
transformational efforts.

Add to them external advisors who have some knowledge of and
as much experience as possible with Electronic Business. Consultants
may be helpful in this context. Many of the traditional consulting
firms (Andersen Consulting, McKinsey, etc.) have rushed to pull
together Internet-based practices over the past five years. Note that
none of them, perhaps with the exception of the two companies mentioned above, foresaw this explosion in Electronic Business. Andersen Consulting, however, still emphasizes Internet services although
the market has passed on to Electronic Business. Recognize that
much of what they will provide is theory and unproven techniques.

External advice can also come from companies that have experience with operating Electronic Business already. Companies such as
Hewlett Packard, IBM, Siemens Business Services and Sterling
Commerce have designed and built operating Electronic Businesses
around the world.

Siemens Business Services has developed a series of what it calls e-
SPEED service packages to take companies to a complete Electronic
Business solution. Each of these packages comprises modules that
can be customized to match specific business requirements. Sectors
addressed by the e-SPEED services include financial services, industrial, public sector, telecommunications, transportation and utilities.

IBM, of course, was the first company to really recognize the importance of Electronic Business. The company grabbed the thought
leadership of the industry in a way that it had not done for 30 years.
Its opportunity and problem now is the implementation of its Electronic Business processes and living up to the promises it has made.
IBM itself is a living example of a company that is transitioning to an
Electronic Business model. IBM CEO Lou Gerstner is driving this
transition but he did not identify it. Others fairly far down in the organization (then) identified the coming Internet explosion as early as
1994 (John Patrick) and the Electronic Business potential in 1996
(John Thompson, Irving Wladafsky Berger). What is most impressive
is the speed with which IBM embraced and implemented the con-

cept: Elephants can dance! Hewlett-Packard, Siemens and Unisys have quickly followed.

This is an important point when talking to any external organization about Electronic Business. How has it changed itself because of Electronic Business? If it has not changed, do not use it. If it "talks the talk," it must "walk the walk"!

Organizational Barriers

The changes that Electronic Business both brings about and makes possible will transform an organization's structure. Focus must be dual: on what will be offered to customers and suppliers outside the organization as well as on what new means of operation will be put in place internally. Larger organizations can use the opportunities offered to sharpen response times, trim infrastructure and encourage the abilities needed to compete with smaller organizations. Smaller concerns can self-organize to take advantage of the wealth of Electronic Business tools and processes that level the barriers of entry, allowing them to compete against giants and across global boundaries.

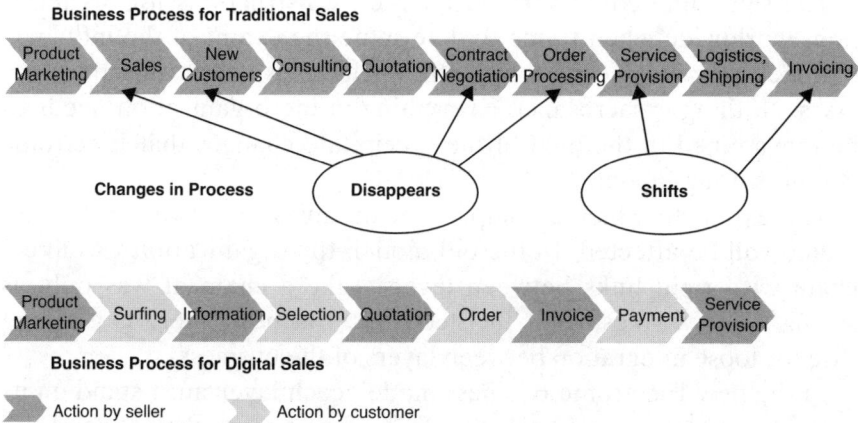

Business Process for Traditional Sales

| Product Marketing | Sales | New Customers | Consulting | Quotation | Contract Negotiation | Order Processing | Service Provision | Logistics, Shipping | Invoicing |

Changes in Process   (Disappears)   (Shifts)

| Product Marketing | Surfing | Information | Selection | Quotation | Order | Invoice | Payment | Service Provision |

Business Process for Digital Sales

Action by seller   Action by customer

Figure 3. Change in business processes as exemplified by sales processes

Fundamental questions relate to geographic structures, sales and distribution channels, customer support facilities and locations, etc. In looking at such structures, always remember that Electronic Business is non-geographic and strongly cultural.

Reluctance or outright resistance to Electronic Business organizational restructuring can be a great barrier; people do not change unless they have to. The political and business impacts are so severe that many people will not make the transition. In fact, many managers will adopt the "not on my watch" syndrome. This syndrome originally referred to naval officers who tried to avoid being on watch when their ship came into port as this was the time of maximum danger. Executives in many companies are the same: They will set up studies and pilots and make small moves as they look forward to their retirement in a few years, putting off "coming into port" until after that time. Then their successors will have to deal with the real problems.

Impact on Organizations

In his effort to rescue Chrysler, the US automobile giant, from a sales slump, its chairman, Lee Iacocca went on American television in a commercial to promote the company's redesigned line of cars. In the commercial, Iacocca used a tag line that later became the rallying cry of the emergence of the Internet throughout the world: "THIS CHANGES EVERYTHING!"

The same line could very well serve as instructions for organizations worldwide when trying to determine the extent of the influence of Electronic Business on operating principles, structure and processes. Nothing is sacrosanct. Everything in the organization needs to be reexamined in the light of the precipitate changes that Electronic Business brings about.

The extent to which a company or organization is vertically integrated will be affected. In the old model, the organization was like a chain with rigid links between levels. Like a chain, it was only as strong as its weakest link. In Electronic Business, there is a strong case for loose integration between layers of the business.

In the new Electronic Business model, each layer must stand on its own. It must be "market-tested." Instead of a rigid chain, each level can operate semi-autonomously; it can form independent market relationships that often compete with other levels within the organization. Instead of a chain, we have a network of loosely coupled organizations. This provides greater flexibility and overall strength. It enables the Electronic Business to adapt to changes in market structures and performance of individual layers or competitors. It enables the Electronic Business to focus on its core competencies, as the BPR experts would suggest, but it also enables the Electronic Business to experiment and to adapt relationships to cultural and other variations in the Electronic Business world.

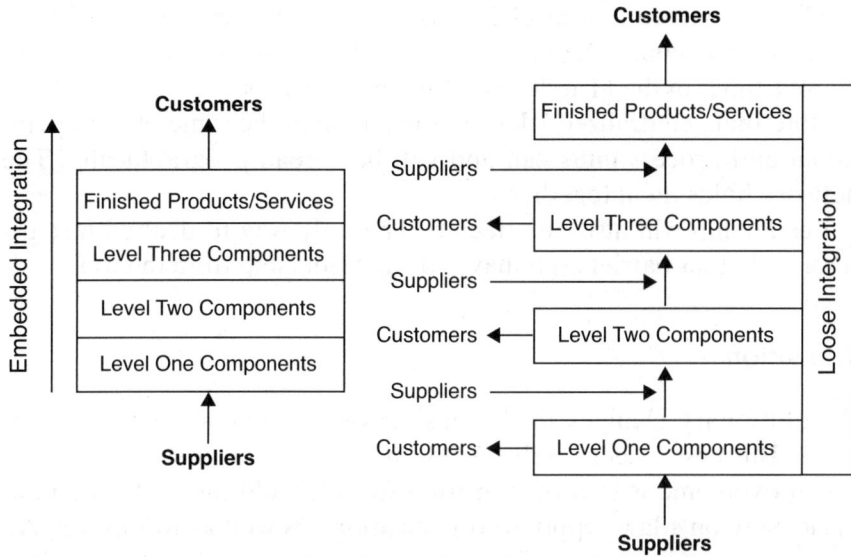

Figure 4. Traditional vertical organization is tightly integrated (left); Electronic Business organization is loosely integrated

Computer systems companies such as Siemens and IBM provide good examples. Historically, these companies made semiconductors for their own use in computers. Their software only ran on their computers. Their peripherals would only attach to their computers. Their services only supported the sale and operation of their systems. In the Electronic Business world, they have been transformed into networked organizations. Each layer can deal with other companies, not just its own equivalent unit. Thus, even IBM's software is now being written to run on non-IBM platforms.

Despite the recommended network of loosely couple organizations, the overall organization does not become a holding company with a set of autonomous organizations. Far from it. The core message and vision of the organization is strengthened. Units must follow that vision and have a viable place in it; otherwise, they are transferred out of the organization usually as part of a market-strengthening activity. IBM's arrangement with AT&T to handle IBM's ISP activities is an example.

In this Electronic Business environment, processes must be examined as well as business units themselves. Thus, "cash" collection for example may be a process that is "centralized" across units and market-tested. If the internal unit cannot provide a competitive service, it can be outsourced to a specialist organization. If it is market competitive, it can become a commercial unit itself. This has happened numerous times in the IT industry with mixed success.

Note that "centralized" does not mean all in the same place; operational and process units can and will be spread geographically. The network holds them together.

Leadership will also be affected. The only way to deal with structural and other barriers is to have strong leadership from the top.

Transition

Revolutionary changes in the business environment, such as Electronic Business, occur rarely. When they do, one of the biggest barriers to overcome is that of transition from the old model to the new. Isaac Newton's laws apply to organizations as well as to physics. An organization will continue in its state of motion until and unless it is acted upon by an external force. When it does change, the change in motion is directly proportional to the extent of that force.

Electronic Business is a very strong force that will act on most organizations in the near future in Europe. For many organizations, the change in motion will be fairly abrupt in historical terms. That means that changes will occur over periods of a few years as opposed to tens of years. On the other hand, they will not usually occur over days, weeks or months, at least yet. (In the middle of the 21st century, this shorter period of transition will probably become more common.) The timing of changes will vary by industry and location: Some organizations, particularly in IT and information intensive industries, will have to react to market changes in very short periods. Advertising agencies are another good example of organizations with short transition periods.

Thus, each organization must analyze the speed of transition required and its variance by unit, process and location.

The kinds of problems inherent in this kind of transition are demonstrated by changes at a hypothetical retail bank. Consider a physical bank branch with 10,000 customers. The bank introduces a fully electronic bank, of which there are only a few examples at present. Say 1,000 of the customers of this branch transfer to the new electronic bank. That leaves 9,000 that support the existing branch infra-structure, whose costs of operation are only minimally reduced by this change. Clever management and application of new systems can still maintain the profitability of the branch.

Now a further 1,000 people leave the branch for the electronic bank, which now includes video conferencing and higher level electronic banking interfaces. Profitability now becomes problematic for the branch. Certain activities are transferred to advanced ATMs and smart kiosks, but these are almost always part of the electronic as opposed to physical bank. Oh yes, the electronic bank unit is still not profitable as it spends large amounts to develop and compete in the critical new electronic banking space.

At what stage does the bank consider closing the physical branch and transferring all the customers to electronic interface? When it does so, it will lose customers to boutique banks that are set up to handle those local customers that want the physical touch. These may not necessarily be the customers that the bank wants to lose.

Let us not think that only the small, IT dysfunctional customers will want to use the physical bank. One of the busiest financial "branches" we have seen is the Fidelity branch on University Avenue in Palo Alto, Ca. This is right in the heart of "Silicon or Software Valley" and the patrons certainly look as though they are electronically enabled. Will it continue and for how long?

This is the critical issue in transition: How fast do you change over?

What impacts does this scenario have on real estate leases, branch bank construction and location, personnel policies, etc.? And how can banks use their physical locations advantageously?

In a parallel manner, automobile service stations have changed dramatically in the last decade and become retail locations serving far more than gas or petrol. IT systems and technology largely made this change necessary and possible. Electronic diagnostic systems requirements and on-board electronics made expensive automation necessary except for simple tasks like changing tires. This pushed service back to dealers and large service facilities and opened up space for retailing. Electronically enabled dispensing and self-service systems improved productivity and reduced operational costs while increasing capital costs. Credit card processing and POS systems changed administrative productivity. Thus, old stations became un-workable and many closed. There are far fewer service stations and far more cars today than 10 years ago. The industry changed, and the transition has been fairly abrupt.

We are not suggesting that banks become fast-food locations. Or are we? Perhaps they should consider it.

In most cases we will have a Boolean "and" situation rather than an "or." This means that both the physical and electronic markets and requirements will exist. The mix will change with time and location. It will be a requirement to be in both worlds at the same time.

The transition issues inherent in such an organization-wide review and reformulation need to be attended to with the utmost care. Specific transition teams can play an important role, especially once transition issues have been identified and action plans have been settled upon.

How Do You Organize Electronic Business?

A critical decision to be made is how to organize to enter the new environment. Is it better to go forward on such a thorough business transformation using your existing organization or using a "green fields" approach?

The existing organization contains the acquired knowledge and talent that has taken the organization to its current state. But it may drag its feet because of the problems of "eating your own children." And the knowledge it has is rather like understanding manufacturing before the Industrial Revolution. By definition, the new approach will cause power shifts in organizations. Budgets will be changed. Head counts will be altered. Organizations do not change themselves; they have to be changed.

One approach is to set up a brand new organization that is more agile at implementing the required changes. It has no "baggage." The actual costs and benefits as well as the likely outcome are also clearer.

But this approach raises the question of who manages the transfer. Meanwhile, the "old" organization will not stand still. It will take initiatives to bolster its position. It may actively or passively sabotage the new organization. It can cause confusion in the marketplace.

Furthermore, the whole organization does not benefit from the new unit; the market and the investment community regard it as "old," despite its movement into the new space. We have repeatedly seen market values of companies shoot up when they announce that the whole organization is switching to the new Electronic Business model.

Another approach is a combination in which the new organization is responsible for the development and the old responsible for the sales and support of both the old and new models. Although fraught with difficulty, this approach has some merit. It largely depends on how the sales and support organizations and people are rewarded and managed. Each situation is unique.

Some companies solve this problem by spinning off new divisions or subsidiaries, issuing tracking stocks for new entities and the like. This can set up competition with the old business, because the simplest place for the new company to get customers is the "old" business.

Responsibility

A major barrier to change is responsibility. As we have already stated, most executives are loath to make substantial, "bet your company" changes unless and until something happens externally. At INPUT, we have a saying: "Desperation is the mother of business process engineering!"

Even when executives see the necessity for change, their directors, investors and management often hamstring them.

In our view, the only executive that has the power to take the kinds of action and make the kinds of plans required by Electronic Business is the chief executive. In some cases, this will require that the CEO be changed, even if she or he knows what to do, because it will be the team that must change and to do so would require a new leader.

Certainly the CFO, the CIO and the operational executives on the team can support the CEO. They may even be the initiators and/or leaders of the investigation and transition processes. But the leadership must come from the CEO, with the agreement and cooperation of the board.

This requires a substantial investment in research and education so that the issues are understood to the greatest extent possible, consistent with acting in a timely manner. This is the CEO's responsibility to drive this process.

Companies do not have the luxury of years of analysis, nor a well-established body of research to call on.

Neither is this a one-time activity. As we have seen already with Web sites, the Electronic Business path is a slippery slope that can only be successfully negotiated with difficulty. It is an ongoing process that company leaders must commit to for the long run.

Channel Conflict

The cheapest place to buy anything is at the factory gate. The result of this truth has accompanied the rise in the real world—the "bricks-and-mortar world"—of a plethora of manufacturers' outlets. The Internet crosses many barriers to reaching that gate, but raises a whole host of questions in so doing.

The first of the questions is Whose customer is it? Does the customer belong to the manufacturer, the distributor, the dealer, the discounter, the salesperson or the Web site? The previous discussion of Dell Computer highlighted the controversy surrounding the "channel versus direct" debate in the computer industry.

A brief discussion of the automobile industry will further clarify the issues in the channel conflict that the Internet will only exacerbate. In the automobile industry, it will be the manufacturers versus the dealers. Internet-based facilities such as Autobytel will transform the role of the auto dealer and the dealer's relationship to the manufacturer.

Autobytel.com, and other Internet-based services like it, are information compendiums and ordering mechanisms. As discussed at length in Chapter 1, these Web sites aggregate information about autos, allowing potential buyers a place to do "one-stop shopping." Such Web sites serve both consumers seeking an individual car or truck as well as business people seeking fleets of vehicles.

This kind of service promises a number of internal fights over customers in the future. If the manufacturer can sell a car and arrange to have it delivered, what role will be left for the dealership? Will they become display spaces for auto models or something more? They will certainly serve the used-car market for some time to come, but even that market is experiencing consolidation by acquisition as well as an enlarged role for Internet-based companies.

Situations involving in-house conflict can also occur. When a company decides to implement direct sales but maintain its dealer channel, what happens to the internal groups within the company that were devoted to servicing the dealer channel? Are they put in competition with the direct sales people who were selling the same products or services to customers who previously would have moved through the dealer channel? This is the kind of problem Compaq faced in its battle with Dell and its decision to go direct.

Compaq is not alone in this escalating conflict over the channel. This is an "and" not an "or" situation. In the future, companies will find that they must keep the old infrastructure as well as embrace and invest in the new sales channels, such as direct sales and Internet-aided sales mechanisms.

Operational Barriers

Once the organizational imperatives have been identified and prioritized, a set of operational issues specific to Electronic Business present themselves. As noted above, there are many complicated problems associated with the move into Electronic Business territory. Nothing about this kind of shift is simple and can be taken for granted. Below are outlined some of the basic concerns as Electronic Business operations go forward.

Management of Security, Authentification and Access

With an organizational Web site and other resources made available over the World Wide Web 24 hours a day, seven days a week, 52 weeks a year, indefinitely, the security of the data posted on that site, its integrity and its invulnerability are of paramount importance. Since the site may well contain data and information aimed at a variety of audiences, and because some of the contents of the site may be intended only for limited access, authentication plays an important role. Authentication services will be employed to ensure that the right information reaches the right people and not others for whom it is not intended.

A business or organization may have only a few external Web sites, perhaps each revolving around a group or family of services. Larger organizations and companies, however, may have scores if not hundreds of internal sites, each devoted to a particular aspect of operations or business. In some companies there may be a Web site or page for each employee. Here, too, security and authentication issues become of the utmost importance. Information stored within human resources' databases, for example, may be quite sensitive and need to be protected from employees in other parts of the concern. In these situations, access management becomes an organization-wide priority, requiring specific policy initiatives and monitoring. (A discussion of security issues in terms of technology is presented later in this chapter.)

Creation and Maintenance of Content

The data and information residing on an organization's Web site is a direct reflection of that organization and in the future will represent the most public face of the organization. As the organization moves into new areas, perhaps offering new products and services, or updates of existing offerings, the data contained on the Web site will need to reflect those real-world changes.

The creation of the Web site is an important undertaking. Design factors, as well as judgments of exactly what to include and how to protect it from unwanted changes, are critical decisions that require precision and certainty. Insuring that the site reflects the philosophy as well as the products and services of the organization it represents cannot be overemphasized.

Potential employees and customers check an organization's Web site as a matter of course today. They also check competitors, chat groups, etc. Keeping up-to-date therefore becomes vitally important.

Chat groups are particularly difficult to "manage," but will be increasingly important to monitor and manage. Organizations will have to manage communities as well as content.

As discussed previously, the professionals who have recently honed their skills in Web site design and creation are in great demand and may well command very high fees and salaries for their work. This kind of investment may prove to be among the most important an organization undertakes. Critical design and content decisions may well set the course for the development and growth of the organization.

Once the Web site is online and actively available, one of the most important tasks will be to communicate to users that the information they are accessing is the freshest and most current available. Maintenance of content is an important company-wide policy issue. There are programming and coding tricks that can make a Web site appear to be up-to-date, like the macro that automatically posts today's date at the top of the Web site's home page. Sophisticated users and Web surfers who return to a site on a regular basis, however, will soon discover what content is fresh and what content has been left to languish from earlier postings.

The currency of postings on Intranets takes on a new urgency. Intranets will become the central organizing and communications mechanism of an entire company. As the lifeblood of company processes, the Intranet requires content maintenance policies as rigorous as those intended for the Internet Web sites meant for outsiders. The timeliness of postings on an Intranet can determine how quickly or slowly changes in data, policy or other issues are communicated to an organization's entire staff, no matter where that staff is located geographically.

Performance and Scaling

The systems needed to maintain Internet and Intranet systems involve people, equipment, networks and software. Key components that affect performance are shown in Table 2. Failure in any one of them will be unacceptable. Putting these together effectively and economically will be a major barrier to overcome on a continuing basis.

Table 2. Factors affecting performance of Electronic Business resources

Servers	The performance of the server computers used to house the Web site and carry out the Electronic Business processes
Storage	The size and performance of the data storage systems connected to the servers
Switches	The performance of the switches and other connections to the communications infrastructure
Applications & Systems	The performance of the applications and systems software programs used
Networks	The performance of the telecommunications network(s) connecting the various nodes on the network
Support	The availability and performance of the support functions, both human and systems
Integration & Fit	The ability to integrate software with legacy technology

The performance of the whole depends on how well the parts fit together. It also depends on the basic access demographics: who is accessing the system, how often and what for. In planning the "who," we would like to plan the numbers, location, access method and characteristics of the access entities.

Unlike old IT systems, the new Electronic Business systems are characterized by unpredictable traffic patterns. Access characteristics can vary widely. Taking account of the potential peaks and valleys is therefore much harder to do.

The CIO of a major Web company recently told us that his normal (not low) level of operation was 7 percent of his planned peak, and they just had to close down because that peak had been exceeded. Performance counts. One factor that will surely lead to loss of business is the electronic equivalent of the busy signal on the telephone or even worse, the "Your call is very important to us. Please hold on while we deal with other customers" message. Electronic patience is not a characteristic of the Electronic Business environment.

In this context, scaling is one of the most important selection and operational criteria. Scaling is the ability of an Electronic Business system to rapidly adjust to major unanticipated changes. For example, when the Mormon Church put its genealogical database online, it immediately became one of the most accessed sites on the Internet. The system was not designed to handle that much traffic its first day. Yet by the next day, the system had been adjusted to handle the traffic volume. That is scalability!

A major retailer had a similar experience when it went from 50,000 to 500,000 SKUs (stock keeping units) online on its Electronic Business network. In this case, the increase in volume caused the system to break down completely and led the retailer to switch its IT supplier.

In the future, however, suppliers will not have the luxury of even one day to make adjustments of this nature.

One consequence of these needs for performance and scalability is a trend toward outsourcing of Electronic Business IT support infrastructures. IT services companies can rapidly adapt to changes in the Electronic Business environment; they can scale far more rapidly than an in-house organization.

Security policies and procedures also affect performance and scalability. Most people would say that every effort must be made to provide a secure Electronic Business environment. Of course, this is impossible or at least prohibitively expensive.

In the real world, it is possible to have very high-security environments with multiple access controls, security guards and walls 10 feet thick that will survive almost anything up to a direct hit by a nuclear weapon. Yet secrets are still stolen. Every security precaution decreases ease-of-use and inhibits performance. If security considerations make an Electronic Business hard to use, people will not use it unless they have to. Thus, there is a constant trade-off between security on one hand and performance and ease-of-use on the other.

User Interface Barriers

User interfaces must facilitate ease-of-use. To date they have not, primarily because the computer industry has been the purveyor of access systems. Computer systems were built for "techies," not for the general public. This is about to change.

In the first two decades of the Internet, facilities such as e-mail, collaboration and file sharing were the way this new network of networks was put to work. In the early 1990s, Berners-Lee at CERN developed the concept of the World Wide Web, the great Internet content container, based on the hypertext principle. Mosaic, the original Web browser, provided a way to view multimedia content on the Web.

Computer penetration in %

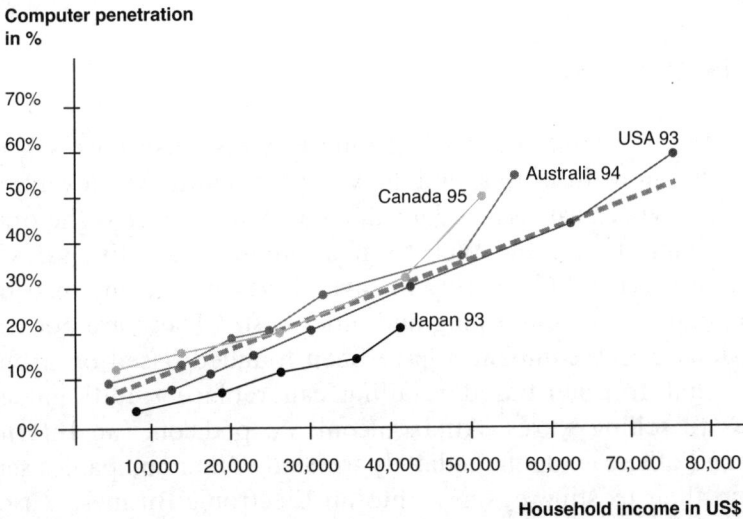

Figure 5. Computer penetration rates as a function of household income (OECD)

The browser is now the universal access tool for the Web and the Internet. Soon, however, new Internet access tools will radically change the very nature of the browser. As devices beyond the PC (personal computer) gain Internet access—for example, telephones, personal organizers and the like—new technologies will transform the browser in profound ways. ASR (automatic speech recognition) will be but the first of these new technologies, to be followed, most likely, by others such as touch pads, gesture recognition and more.

Managing the interface issues for today's browsers is difficult enough. Integrating the support for future browser technologies will present any number of new demands and challenges. But it must be done. The rollout of these new interface techniques will not hesitate or pause. ASR- driven Web devices are already on the market (e.g., Philip's Nino, Qualcomm's pdQ, etc.).

Plan on the Web becoming as easy or easier to use as the TV or the telephone. Several years ago, a survey was carried out by an IT vendor as to which computer interface method was the easiest to use. The winner by far was the game controller interface used by children (and a few adults!) to play electronic games.

Some readers may be familiar with the science fiction book *Ender's Game* by Orson Scott Card, which used as its theme the use of an electronic game to carry out a real war. Fiction often predates reality.

Financial Barriers

Many of the most important Electronic Business decisions will revolve around how much to spend, how quickly and in which ways on what things. What business model will a company adopt? Is the organization aiming to use the Web to make money via active sales of products or services? Or merely to save money by shaving distribution, communication, publishing and other costs? These are not trivial decisions. Some companies have been founded based on an hypothesis that Internet-based retailing can replace traditional approaches to selling wares. Amazon.com, Peapod.com (an Internet grocer) and others come immediately to mind. Other companies seek to extend their existing business into an Electronic Business. Cisco, Dell Computer and IBM have so far been the most successful in converting older business models in this manner.

Costs

The costs of Electronic Business are hard to predict from both developmental and operational viewpoints. Since the environment changes so rapidly, development becomes a process rather than a project. An organization's Electronic Business offering becomes a series of releases rather like episodes in a serial. Many organizations do not know what their costs are in the early stages of Electronic Business development, the "access" and "presence" stages mentioned earlier.

Telecommunication costs affect Electronic Business potential, especially outside the United States. In many countries, telecommunications carriers charge for Internet access by the minute and often impose other costs. In Europe, this situation came to a head late in the spring of 1999, when a protest was organized. Internet users were encouraged to stop both e-mail and Internet surfing for one day as a demonstration to telecommunications concerns that access rates were too high. The effect of the "Net strike" Sunday was roundly debated, but the attitude of one representative of British Telecom was typically insular: "We are constantly reviewing our pricing and listening to our Internet customers, but one way or another customers have to pay for access. We just think our method of charging by usage is fairer."

That thinking has been turned on its head by the "free" (users still have to pay for local phone charges) Internet access services that are springing up, not only in the United Kingdom but elsewhere. BT may act like the emperor with no clothes for a while but eventually the economics of the situation will prevail.

Similarly other companies have tried to introduce "roaming" charges; again these have been met with disaffection to no-roaming–oriented organizations.

Switching an ISP vendor is relatively easy. In fact, most buyers establish relationships with multiple suppliers in order to take advantage of price and availability changes. Some or all of these suppliers can be from countries other than the location of the buyer, provided they can provide local-access points of presence (POPs).

This reduction in switching cost is a fact of life in Electronic Business that must be considered by all organizations. In the physical world, switching providers is often expensive and time-consuming. Price changes of at least 25 percent are often required to get buyers to switch from an existing provider, or a substantial improvement in service or product performance is required. Differentials of at least 10 percent are required to differentiate for a first-time purchase based on price alone.

In the electronic world, switching is far easier. Organizations of all kinds are going to make it easy for the buyer to switch. This will increase the "churn" rate of customers. Loyalty-building programs, which are sophisticated bribes, will become common.

In the United States, telecommunications companies, though freed from much government regulation several years ago, started to impose higher fees when usage of the Internet started to take off. Their efforts to raise pricing or change access pricing structure was met with derision and charges that the companies' own lack of adequate planning had compromised their positions. As Internet usage increased, so did telecommunications usage and US telecommunications companies have seen increased demand for their services. There are also new competitors such as Qwest focus primarily on Internet network access.

As this situation spreads in Europe, access costs will also come down rapidly. If they do not, European business will atrophy or move. Companies do not have to stay in their country of origin.

Pricing

The Internet changes the very nature of pricing. In the real world, it is often difficult if not impossible to effectively compare prices between competitors. Many department stores have been known to send out stealth pricing spies to competing businesses to ascertain what prices are being charged for specific goods and services.

In the online world, pricing is far more transparent. In addition, online services based on providing pricing information, as Priceline.com does for airline tickets and hotel room rates, have recently emerged with a new business model aimed at spurring e-commerce purchases. The assumption is that a user will be more inclined to finalize a purchase decision if presented with complete pricing comparisons all in one place.

This will soon become a much larger business proposition because of the introduction, already under way, of pricing software systems, known as "agents." These software packages go out and surf the Internet and return to the user's computer after gathering all the pricing information that can be found on a specific product or service. The analogy to the search engine spider's scouring of the Internet for information matches is apt.

In a world where complete pricing information is available, purchases are often made specifically based on price alone, as opposed to the consideration of other factors, such as delivery means, reliability, after-purchase care, and the like. Amazon.com, BarnesandNoble.com, and other online book and music sales companies are in very stiff competition in which many times price alone determines a purchasing decision. Discounting works best in commodity businesses such as these.

Billing

The Internet offers a number of solutions to the problems inherent in billing customers. Many companies choose the Internet as the billing medium of choice. One of the first ways in which the Internet was applied to billing was in substituting e-mail for the more conventional posting of bills through postal systems and other delivery mechanisms. E-mail costs, when compared to these other, "snail mail" methods, are considerably cheaper.

A more advanced Internet approach to Electronic Business billing solutions involves arranging for Web sites or Web pages that display information related to the accounts of specific customers. Such systems are usually used in conjunction with e-mail reminders of bills due and their due dates for payment. The Web page system offers the advantage of providing the customer with the most up-to-date information available about the nature and status of any account. Such systems can be linked to existing enterprise accounts receivable systems already in place in legacy computer systems at an organization.

One of the decisions to be made about such billing services is whether to develop such systems in-house, outsource the billing services to a third party, or arrange to have a third party set up, administer or monitor a new Internet-based in-house system.

Collection and Payment

Efforts are being made worldwide to provide secure payment systems that are acceptable across business and institutional structures. Many people in the United States, Europe, Asia and elsewhere routinely use credit cards to make payments for goods, products and services, even across international boundaries. The encryption of that credit card data is often important, but sometimes ignored.

Several companies such as CyberCash, Compaq (Digital Equipment) and others have made attempts to devise payment mechanisms that would facilitate small exchanges of monies over the Internet. To date, none as met with much success. Most recently, credit card companies and issuing banks have teamed with larger companies such as

IBM and America Online to try to come to grips with this problem. Once these issues and other related concerns about encryption are settled, businesspeople and consumers will routinely use the Internet to make purchases.

(The role of encryption and the perception of security are dealt with in more detail in the section on security, later in this chapter. Electronic bill presentment and payment and digital money are addressed in Chapter 6.)

Digital Cash

If at first you don't succeed, try, try again. Banks and other financial institutions have been grappling with the notion of digital cash for several years. No one solution to the problem has emerged. Smart cards, popular in Europe, have come the closest to providing digital forms of money that can be widely used and accepted. Smart cards that can be loaded with new forms of digital cash over the Internet already exist.

The problem has been that no standard has been adopted by all of the same cooperating institutions that routinely send financial transactions around the world. Once again, problems linked to the acceptance of encryption and other standards for ensuring the security of money equivalents stands in the way of standardization and general usage.

Accounting Systems

The power of the Internet to collapse and obliterate national boundaries and international barriers to markets carries with it a series of other problems. Accounting procedures, rules and regulations are often the result of years of intense negotiations. Electronic Business makes its own demands on accounting, not the least of which is how to treat income from sales that take place in the virtual state of cyberspace. When and where do they occur? How do you handle the issues of returns? What happens when you take the "float" out of the system?

There are two basic requirements for accounting systems for Electronic Business: They must measure the right things and they must measure them at the right times. Also, accounting systems serve two audiences: employees (including "managers") and external interested bodies, such as investors.

For external review purposes, Electronic Business accounting problems relate to both balance sheet and income statements. How do we take account of the value of customer bases? A company with no revenues but 10 million "customers" was recently acquired for several hundred million dollars. How do we value intellectual property? Software companies with no revenues are being sold for tens or hundreds of millions of dollars because of their technology.

Existing accounting systems do not account for this adequately; our systems are built for the industrial age. The very timing of reports makes them of decreasing value in an era of very rapid change.

Even attempts to address the problem by institutions such as the Centre for Tomorrow's Company in the United Kingdom deal primarily with restructuring annual reports that are obsolete and often useless when they appear.

How also to deal with cash flow models that are inherently unstable? Accounting systems today are built on deterministic principles. In the future, they will have to be far more heuristic and use modeling similar to the Black-Scholes risk models in the investment community.

Taxes

The United States has taken the lead here, with its national government endorsing policies that prohibit extra and added Internet taxes.

European, Asian and other government entities have, in many instances, taken a wait-and-see attitude about how to assess taxation policies where the Internet and other Electronic Business issues are concerned.

But taxation policies will change over time, especially when governments and legislators begin to grasp the sheer economic power that Electronic Business makes possible. As revenues in the Electronic Business domain continue to expand at very high multiple rates, new taxes may well be put in force. Issues of double taxation are bound to arise as the vexing notion of virtual space becomes intertwined with revenue generation and collection. For example, sales made via a Web site located in the Caribbean to a customer in an airplane, selling goods developed and manufactured in France, but sold by a Taiwanese company presents an impressive conundrum.

The problem today is that we do not know what future taxation processes will be. Yet we set up institutions with an expectation that the situation will continue as it is.

Technology Barriers

Security

Recent surveys have shown that security issues are viewed differently in important ways in different geographic locales. US respondents do not believe that security is an inhibiting factor in e-commerce. In the United States, respondents rate security as 4.5 in importance on a 1-to-5 scale. The satisfaction rating for security, however, is 4.2, which shows that the respondents believe that security is under control. Many respondents commented that security is more of a perception issue than a real issue. Certainly, vendors of e-commerce services and equipment must ensure that their products contain security protection and are compatible with popular security products and technology.

European respondents have a different view of security and e-commerce. They are more concerned about security than their US counterparts and are less satisfied with security products.

Some 44 percent of European respondents consider security to be an inhibiting factor for e-commerce, in contrast to only 14 percent for US respondents. The lag between the United States and Europe is the reason; the response to US surveys in 1996 and 1997 were very similar to the current European one. It can be expected that as more progress is made on worldwide security standards, future European attitudes will reach levels analogous to contemporary US views.

For the first time in a survey of this type, some people actually felt they could get a higher level of security from Internet-enabled systems than they could otherwise. Benefits quoted included the following:

- Security, confidentiality;
- Security of trading;
- Speed and security; and
- Security.

But Internet security remains a large negative for some:

"We aren't the least bit interested in the Internet. Apart from having a Web site, we don't use it at all—lack of security is the main reason," said one $150 million US-based manufacturer when asked about the Internet and security.

The security gap will close as more is invested in Internet security products, services and management. It probably will not close completely because there is always a gap between what users want and what they are willing to pay for.

Perhaps more significant are the gaps in the access to data and ease-of-use characteristics. As stated earlier, the Internet provides much wider access to a more diverse audience. Therefore, the tools and techniques used in the tightly controlled environments of the past will be unlikely to work effectively in this new environment.

One of the most fascinating aspects of the rise of the Internet has been the treatment of security issues in the international media. Since many people around the world do not make use of the Internet in any way, and many who do use it are particularly reluctant to employ it when any security issue is in question, it is surprising to see the aggressive treatment of security lapses over the Internet.

In the mid-1990s, as the Internet was just beginning to be an important social and commercial force, the media were widely reporting on the possibility that hackers and other computer literate malcontents would routinely seize credit card data and other important information as it traveled across the Internet. Many stories emphasized that this was made even more likely since no one was in control of the Internet, and no one owned it or would accept responsibility for losses that occurred by way of Internet-based transactions.

Respected media outlets, particularly news services such Reuters and The Associated Press, and well-known newspapers and television systems have jumped on any story connected with the breach of security over the Internet and made it front-page news. These stories, relating to computer glitches, viruses, shutdowns, failures, even the theft of credit card information, have been trumpeted loudly and vociferously. This is particularly true of the UK press, which has exhibited the most strident response to the Internet security issue.

However, the scarcity of security problem stories is noteworthy. Despite, or perhaps because of, the negative media coverage in the mid-1990s, the number of proven Internet security breaches and losses has been very small. What's more, follow-up stories describing what was done to solve these problems have been the rule and not the exception. Great detail has been provided about how systems have been brought back online, how criminals and data thieves have been tracked down and brought to justice, how computer experts have coped with viruses and worms, and where fixes for such difficulties are available on the Internet.

This information has been reported in general interest publications and media outlets, not only in specialty and technical communications media. The result of this reporting, and the speed with which it is produced, has gone a long way toward reassuring the public. This reassurance is not that everything is completely safe; far from it. The general tenor of these reports is that things have gone wrong, and then they have been fixed, but it may happen again.

The general public's perception has changed, as has the reporting of these difficult situations. The media are no longer filled with Cassandra-like warnings of the dangers of going online. More people go online on a regular basis. Since the magnitude of the growth of the Internet is also routinely and broadly reported, the expectation among the public is that, though not without risk, the online world is relatively safe. If something does go awry, it will be noted publicly and quickly.

Contributing to this sense of confidence has been the increase in the value of companies doing business on the Internet. Because of soaring stock prices for Internet companies as well as reports of the strong online components of traditional businesses, the Internet has taken its place in the mainstream. When one of these Internet companies experiences a difficulty, this too is widely and quickly noted in the media. The service outages at companies such as America Online and eBay are important from a financial point of view because of all the money invested in these companies. Such broad financial participation in Internet-related aspects of the economy has added to the sense of security that is now experienced about most things related to the Internet.

There has been a general realization in the US that the Internet is potentially very secure; after all, the National Security Agency uses it. You can get any level of security that you wish. So security will be a concern, but not a barrier to Electronic Business in the near future. Though these feelings and attitudes are most prevalent in the United States, they are spreading quickly to people in Europe, Asia and elsewhere.

Standards

The full promise of Electronic Business will not be realized until buyers, both individuals and businesses, can easily locate and manipulate all of the data they need to make purchase decisions. Though everyday such facility moves closer, there are several obstacles to automatically exchanging data between applications in a meaningful manner:

Currently, the business-to-business Electronic Business applications market is very crowded, with product introductions from established firms such as IBM, Baan, Microsoft, Netscape, SAP and Siemens, as well as smaller, newer players including Ariba, Open Market, CommerceOne, TRADEX, BroadVision, Connect, and Inter-World. Each vendor uses a mix of different database technologies and software development methodologies that hinder open communication between companies.

The user community has a variety of different business rules, processes and product definitions for Electronic Business applications. In most cases, users define and maintain data in a manner that is unique to each organization, thus limiting inter-company data exchange.

True interoperability will require extensive agreement on standardized processes, message formats and data definitions before data exchange can occur efficiently between companies. There are several different approaches to establishing communications standards:

- **Process-specific**. These efforts focus on a particular business process or function. An example is Open Buying on the Internet (OBI), which is a collection of processes and technical standards designed to facilitate Web-based procurement transactions between companies.
- **Industry groups**. Companies within specific vertical industries are joining together to improve Electronic Business processes between firms. A well-known effort is RosettaNet, which is standardizing business processes within the IT industry with the goal of enhancing supply chain automation. Most recently, government agencies in the United States have reached out to embrace some of the RosettaNet processes, suggesting that future compliance in broad areas can be expected to increase in both the business and governmental sectors.
- **Technical standards**. A good example is XML (eXtensible Markup Language), which is a standardized text format that is designed specifically for transmitting structured data between Web applications such as electronic catalogs. XML provides a uniform and standardized way to add data to existing Web data. XML adds on to HTML (hypertext markup language), the Web's main coding language, and will allow new and important information, too often now found only in enterprise database systems, to make its way to the Internet.

Integration

Vendors and users are often trying to integrate their software with legacy technology that is 20–30 years old. This slows down the implementation process. The high level of customization of solutions also makes integration more difficult—70 percent of companies use either customized packaged software or a purely custom-developed solution.

Software developers are attuned to the need to make the migration path from existing legacy and enterprise systems to Internet-ready systems a priority. Software companies are scrambling to develop turnkey systems to provide the migration paths necessary to assure such Internet-ready integration.

The enterprise resource planning (ERP) companies are the most involved from an internal process viewpoint. But new vendors that focus on the external Electronic Business opportunities are springing up everywhere.

An issue for organizations is the extent to which their existing systems can be adapted to the new Electronic Business environment. An alternative would be to move to a completely new IT structure that starts with a "green fields" approach.

The Y2K Situation

All of this has been exacerbated this year (1999) by the Year 2000 computer coding problem, or "Y2K." Because of this problem, resources have not been available to move businesses forward toward Electronic Business goals. The lack of progress on new projects is a direct result of monies diverted from these budgets to handle Y2K issues. One of the benefits of Y2K, however, has been an increased emphasis on testing business systems against potential problems and failures.

Here again, Electronic Business works. Transfer of work to India, the Philippines, Ireland and other countries has been especially popular for English-speaking countries. Russia, with its large, well-educated population of scientists, mathematicians and engineers, is a favorite destination for outsourced programming and mathematics-based work. Other Eastern European countries are enjoying similar growth in outsourcing. The Internet breeds a global community.

Language Barriers

There are very special language problems in the multilingual European markets, where narrow linguistic boundaries must be attended to in order to win and keep customers. This issue is more serious when dealing with individuals, consumers and business-to-consumer issues. But keep in mind that businesses are made up of individuals. Though English is the de facto language of many in the computer-using community, Web sites and other Electronic Business tools and services must communicate with customers and prospects in their native languages in order to reach the broadest and widest possible markets.

Present solutions are expensive and difficult to fully deploy. Many Electronic Businesses offer Web sites with language options for the display of information. But it is not just a question of language; form and structure are also important in dealing with cultural issues. As we have said, the Internet amplifies the importance of culture.

Europe has a particularly difficult problem to deal with in its history of cultural, racial and class barriers. One consequence of a need to be perceived to be a-geographic (non-geographically biased) and culturally neutral is that European companies are adopting a "com" address instead of their country designator such as "co.uk" or "co.de."

These problems will be addressed in coming years by the perfection and implementation of simultaneous translation and transformation technology now being developed. Technology under development in Japan, Germany and the United States, for example, is aimed at providing simultaneous translation services in telephone calls, but new research initiatives have begun to apply similar solutions to the Internet arena.

The problem is also being attacked in other ways. Japanese companies use graphical representations of information, in place of words, as do many companies with progressive design philosophies. The use of graphic symbols can go a long way toward bridging diverse language needs.

Legal Barriers

Multinational corporations maintain legal departments to grapple with the issues presented by multinational business operations. But in Electronic Business, the barriers to creating a corporation with multinational operations are lower than in the past. With smaller firms, even virtual companies, arising on a daily basis to take advantage of the Internet and Electronic Business opportunities, legal problems are bound to provide local, national and international areas of concern.

Here, too, Electronic Business can provide some solutions. Internet publishing enables legal information to be made more widely available and easier to access. In addition, more companies in the Electronic Business space seek to outsource the fulfillment of certain services. As legal problems related to Electronic Business arise and gain importance, virtual legal assistance companies will form to address them.

Copyrights and Trademarks

The Internet's tendency to trample and often obliterate national boundaries, plays havoc with copyright and trademark protection schemes. Many of these rights and protections have been won after years of costly effort, only to be swept away in a hurricane of Web activity.

Copyright and trademark protection varies not only in structure but also in enforcement from country to country. It has been estimated that 90 percent of the software in China is pirated. If this is the "oil" of the Electronic Business era, then protection of IPR will become an ever-thornier issue. Another aspect of this is security, to which we shall return later.

Many companies are producing tools and services that obviate copyright and other protections. This is the old business of producing locks and keys on one hand and picklocks on the other. Other companies provide the facilities that make circumnavigating protections possible. There already have been lawsuits related to this. The Norwegian public prosecutor, for example, has been notified that a Norwegian company is making available software that specifically looks for music files that contain pirated material.

Licensing

Most licensing agreements in force today are based on terms that were negotiated and set into place in the BI (Before Internet) era. Now, many of these same agreements will have to be renegotiated and made to conform to the new realities of the AI (After Internet) era.

One particular case is illustrative of many of the problems in this area. Amazon.com, a leader in Internet retailing, began as a bookstore. As the company grew, it decided to open overseas branches, including an Amazon.com Web site in England. When *Harry Potter and the Sorcerer's Stone*, a book written by J. K. Rowling, a UK citizen, became a runaway bestseller in the United States, Scholastic, its American publisher, was delighted. But when avid American readers, eager to be among the first to get their hands on the next book in the series, *Harry Potter and the Chamber of Secrets,* flocked to the Amazon UK Web site to buy the British edition and have it shipped to them in the United States, Scholastic cried foul. The only solution was for Scholastic to move up its US publication date for the new book. Lost sales were never recovered. This is but the first of many such problems.

Government Regulations

As a result of changes wrought by the Internet and Electronic Business, the minefields of governmental regulations will increase in complexity. When governments try to make corrections based on contemporary realities, or to align with other governments in the new global environment, things often go awry, even with the best of intentions.

An example was the work of the European Union in crafting regulations to safeguard individual privacy rights in the Internet era. In late 1998, the EU put in force extremely rigorous regulations that stipulated that any company, organization or other entity that was involved in commerce or other intercourse with EU citizens must insure that they meet these privacy rules. The EU was especially concerned with the use of data collections from, for example, Internet transactions and similar commerce.

The US government has taken a different route, looking wholly toward voluntary efforts in the privacy arena. The expectation is that as more companies do business on the Web, competition will spur compliance in a more satisfactory manner than regulation. Competitors will pressure providers to conform to uniform privacy regulations and to prominently post their privacy policies on their Web sites.

The end result was weeks of intense negotiations between government regulators on both sides of the Atlantic Ocean as well as counterparts responsible for privacy policy in commercial concerns. Though many US companies do post their privacy policies on their sites and state their protection policies for the data collected on their sites, not all sites meet the stringent criteria demanded by the EU regulations. Also, without enforcement capabilities, many European companies post policies that they do not enforce.

When the EU privacy protections went into effect, in the autumn of 1998, many expected that connections between many cross-Atlantic Web sites would be disrupted. It was not clear what would happen with international commerce conducted over the Internet between EU-based Web sites and US purchasers and vice versa. The end result was that nothing happened: The connections remained in place and, at least for the time being, EU privacy laws were not enforced on US Web sites.

Laws without the power of enforcement are counter-productive. But the threat of legal action—or the fear thereof—can be a powerful deterrent to Electronic Business. It can also be used as an excuse not to act, and that is the greatest barrier Electronic Business must overcome.

5 Industry Examples for the 21st Century

The Internet and high-speed communications easily available everywhere will change the world forever. In the next two years, the embryo will develop. By the early 2000s, new business models will have emerged and industries will be changing rapidly. This chapter presents forecasts for the growth of Electronic Business, including electronic commerce and electronic services to set the stage.

Electronic commerce is the trading of goods but not services. It is the province of the manufacturing and wholesale distribution industries. Detailed discussion of electronic commerce is presented elsewhere in this book.

The primary focus of the chapter is on service industries, especially those associated with money, such as banking. Money affects all industries and everybody, not just banks, so the subjects of electronic payments and digital (electronic) money are covered in detail.

Electronic Business Market Forecasts

Electronic Business will account for over $2.4 trillion in value of goods and services bought and sold by businesses and individuals in 2003. This may double by 2010.

The amount of world trade of products done electronically (electronic commerce) will by itself reach into the trillions of dollars in the early 2000s. John Chambers of Cisco has said that the total amount of world trade done electronically will approach $2 trillion by 2003. He chastised analysts who, in his opinion, were far too conservative. We tend to agree with Chambers. Certainly the amount of world trade done through electronic commerce in 2003 will exceed $1 trillion. It may reach $2 trillion. It is possible that it could reach $3 trillion.

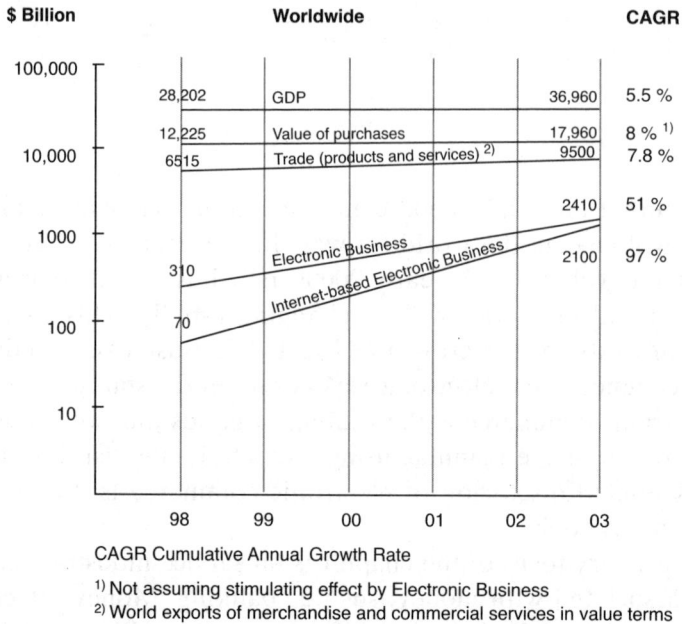

CAGR Cumulative Annual Growth Rate

[1] Not assuming stimulating effect by Electronic Business
[2] World exports of merchandise and commercial services in value terms

Figure 1. Forecast of worldwide GDP, purchases and Electronic Business (INPUT, Siemens)

One factor in this wide band of potential is the degree to which Electronic Business itself will stimulate world trade. All the forecasts of trade done to date do not include the stimulative influence of Electronic Business. One reason, of course, is that there is no way to measure it.

In our opinion, Electronic Business will accelerate the overall growth of world trade, particularly trans-border electronic commerce. This will have a huge impact on shipping and transportation of all kinds. Electronic retailing (BTC) will have a large positive effect on the package distribution and local distribution industries.

Figures 2 and 3 show the expected growth of sales of products and services in both business-to-consumer and business-to-business sectors.

Market Size ($ Billion)

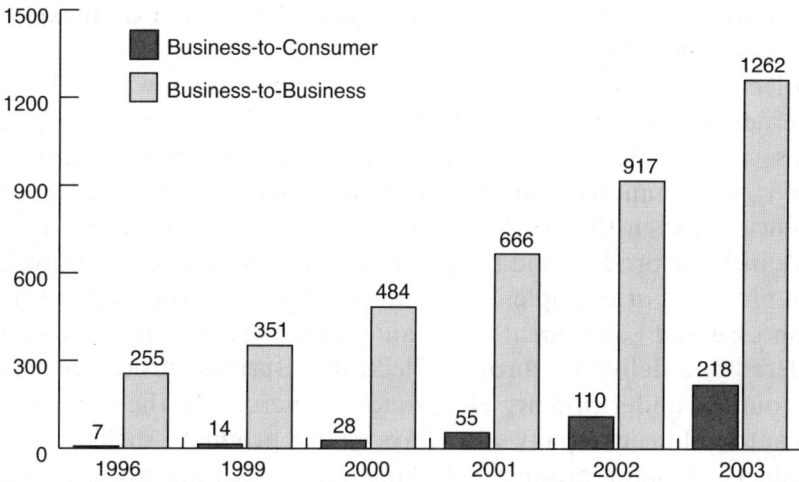

Figure 2. Worldwide value of goods bought electronically (electronic commerce and electronic retailing) (INPUT)

Value ($ Billion)

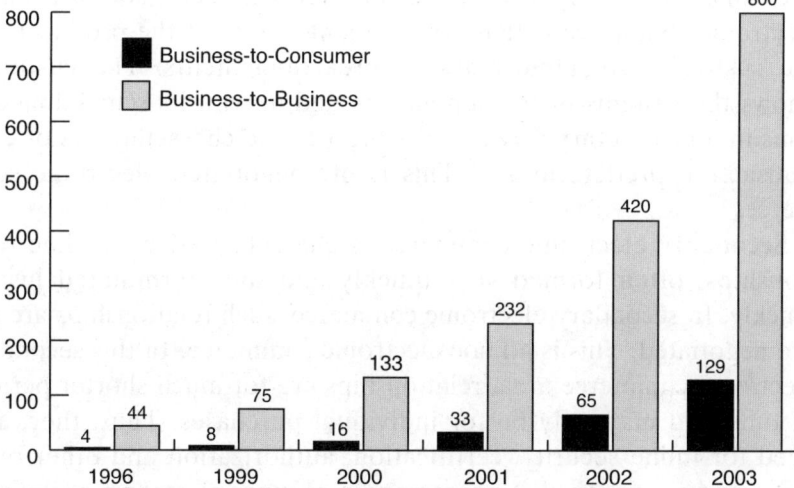

Figure 3. Worldwide value of services bought electronically (INPUT)

By 2003, the ratio of business-to-business compared to business-to-consumer sales will have fallen to less than 6:1 from the 1998 ratio of nearly 40:1. Will this ratio ever become 1:1? Not likely, given the fact that goods can be sold repeatedly in the business-to-business market as they make their way along the value chain. These goods can be sold only once by business to the consumer, however. That makes the 1:1 ratio impossible.

Electronic commerce forecasts can be sliced many ways. There is Internet versus non-Internet, EDI versus non-EDI, and primary versus secondary electronic commerce. The difference between primary electronic commerce and secondary electronic commerce is the difference between the goods that are necessary for a business to produce its basic product and the goods that are ancillary to that production process. For example, in automotive manufacturing, a drive train is an essential component of an automobile. Therefore drive trains ordered and delivered through Electronic Business processes would be counted under primary electronic commerce. On the other hand, an automobile company also buys many products such as office products, cleaning products, clothing, etc., which are not concerned directly with the production of automobiles or trucks. Such items fall into the category of secondary electronic commerce.

Primary electronic commerce has been with us for many years, particularly through the use of closed EDI systems. Secondary electronic commerce is relatively new. It has really only made its appearance since the emergence of the Internet. Characteristic of primary electronic commerce is that arrangements between the producer and the customer are almost always prearrangements. The customer knows the systems of the supplier, is tied in with its scheduling systems in a proprietary way and has most of the characteristics of each transaction predetermined. This is pre-negotiated electronic commerce.

Secondary electronic commerce is characterized by ad hoc relationships, often formed very quickly and then terminated just as quickly. In secondary electronic commerce, such relationships are not pre-negotiated. This is ad hoc electronic commerce. In this secondary electronic commerce area, relationships are for much shorter periods of time and may only be for individual purchases. Thus, there is a need for inline security, certification, authorization and other processes that are relatively unnecessary in primary electronic commerce, in which the supplier is well known to the customer and vice versa.

Internet commerce is just getting started in business-to-business. The primary barriers to growth of Internet commerce are in very large organizations that have investments in proprietary networks. These organizations include automobile companies, aerospace companies, etc. that are loath to change to Internet commerce from their existing systems. Once they do, however, growth will be very rapid.

Europe will grow faster than the United States because it is starting from a much lower base. Moreover, the euro is fundamentally altering the mindset in Europe. Suddenly, companies recognize that not only are there no customs barriers between Manchester and Milan, but also there are no currency barriers. This impact on the business mind in Europe will take several years to fully develop, but there are already signs of it in business press reports of activities across borders. There are no tangible borders left in Europe, either physical or electronic. The only borders are mental and intangible ones such as language and culture.

As Internet commerce applications become more sophisticated and the limitations of traditional EDI more apparent, the mix of Internet- versus EDI-based e-commerce will change, with Internet commerce surpassing the older form for the first time in 2001, as shown in Figure 4.

Market Size ($ Billion)

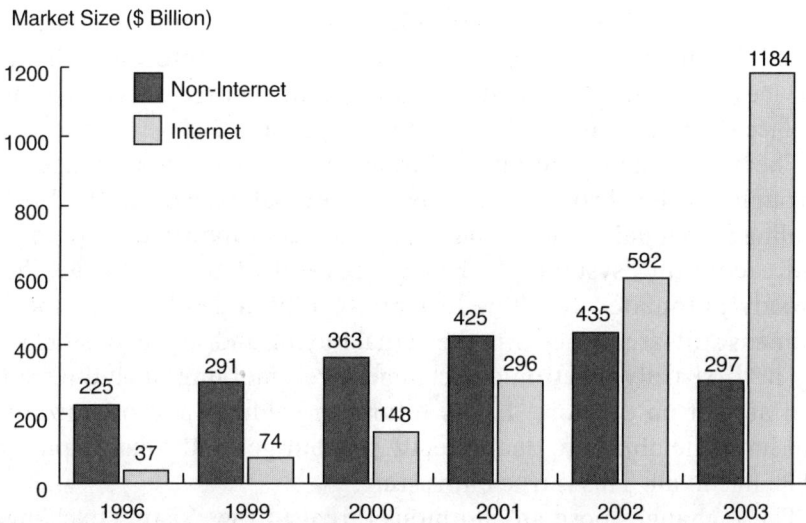

Figure 4. Worldwide value of Internet-based versus non-Internet-based electronic commerce (INPUT)

In services, travel and electronic government will be the major components of the market. The e-government component does not include transfer payments or taxes but only the costs that governments charge citizens for services such as licenses. In this regard government is a business.

Electronic Business outsourcing is one of the emerging facilitating services. Companies such as Sterling Commerce have already signed the first contracts. As companies seek to expand quickly across geographic boundaries, they will find that outsourcing is a viable alternative to trying to set up cross-border organizations and organizational relationships. Although the boundaries are intangible rather than physical these days, some of them are quite vexing (taxation and government regulation issues that vary from country to country, for example). Outsourcing relationships to organizations that can deal with these issues and with the multitude of telecommunications options becoming available in Europe will be a viable alternative to doing it oneself, particularly for startup companies.

Electronic Investing and Finance

In banking and finance, the securities industry sector is the first area that is moving into Electronic Business. At present, the kind of services offered by Schwab, Ameritrade and others are not true Internet-based Electronic Business services. They are "Internet-enhanced services." In the United States, these e-brokerage services simply replace the telephone or physical ordering process.

There is a chain from buyer through a broker to an exchange then through another broker to a seller. In several exchanges the broker dealing is manual, done through floor dealers supported by reporting and accounting systems. In Europe several of the exchanges have already automated to allow brokers to buy and sell electronically. Humans still largely perform the actual buying and selling of stocks.

The first truly electronic exchanges are emerging, including some that are set up offshore. In the future, individuals and organizations will have the ability to deal directly without going through a physical exchange at all. This is true e-investing.

These changes pose an enormous threat to the existing exchanges. One response of threatened organizations is always consolidation. There is a significant amount of that these days in the stock exchange business; witness the NASDAQ acquiring the American Stock Exchange.

Another result of electronic trading is the movement toward 24 x 7 x 52 trading; i.e., 24 hours a day, seven days a week, 52 weeks a year. This is not just for individuals or consumers, but also for corporate treasurers. The really significant money movements today are not those by governments or individuals, but rather those by corporations. It is treasurers who affect exchange rates, more than people such as George Soros.

Electronic investing will lead to much higher volumes and volatility in individual equity and loan instruments. Thus, the pace of change in dealing, trading and exchange processes will accelerate.

Electronic Banking

John McCoy, chief executive of the New York–based BankOne, the fourth largest US bank, has said, "If I'm right about the Internet, that is how we will get our scale!" He is stopping acquisitions that have formed the backbone of the growth of BankOne in favor of investment in the Internet.

By 2000, 40 percent of all banking and finance transactions will be performed online. Moreover, these online transactions will probably represent by far the majority of transactions in terms of value.

By 2002, 90 percent of the major international banks will have installed electronic bill presentment and payment systems. This is supported by research from CheckFree, which predicted in their last annual report that eight of their top 10 banking clients would promote Internet-based bill presentment and payment before the end of June 1999.

Digital Money

Digital money is a concept, technology and/or instrument that is in the early stages of adoption. The inflection between early adoption to significant use will come in the early '00s. Its use will spread extremely rapidly. Physical cash will probably never disappear completely. If it does, it will take decades to do so. Indeed, with "smart paper," we may see a merging of cash and smart card concepts. After all, cash is simply stored value.

But what is digital money? It may be defined as follows:

- A form of money that stores value as sequences of encrypted digits in computer code. This limited, stored value is depleted when a money transaction is completed. Like physical currency, it is transferable and can be used only once in any given transaction. Also like physical currency, values of digital money are divisible into smaller units.

- Typically, digital money has no physical form. It exists entirely as software and is created in connection with a bank deposit account that holds ordinary currency. It is "spent" when a transaction with a vendor who accepts the value specified redeems it from the authorized source, usually a bank holding a local currency demand account.
- Like a paper traveler's check, digital money exists independent of a particular transaction, can be saved or stored, preserves anonymity between buyer and seller, and represents a fixed value (at least initially) in a specific currency. Also like a traveler's check, it can be spent in place of ordinary currency in numerous, if not all, business transactions that require the exchange of money. Because the instrument is secure, the bearer does not require identification.
- Digital money transactions may be conducted electronically through an Internet (or other) computer link, or by means of "smart cards" that capture and record amounts of digital money. The values stored on such cards can usually be replenished when linked to a bank's demand deposit account of currency through an ATM, PC or other system at home, work or elsewhere.

Despite the growing media theme that digital money represents the "money of the future," bankers are deferring investment in this future because they:

- See only mild interest in digital money products at present among either customers or merchants; also, they lack any effective consensus concerning the pricing of digital money products.
- Suspect that many vendors of digital money products are more competitors than partners; they remain convinced that digital money products will enhance their competitive positions, but retain a low level of confidence in vendors.
- Dislike the commoditization of their products and services that can result from close cooperation with vendors of digital money products and/or financial transaction processing.
- Debate and question the whole concept of digital money, smart cards and other approaches to new cash instruments. Many question the efficiency and effectiveness of stored-value smart cards, even as extensions of the traditional credit card business. The concept calls for them to become the equivalent of the electronic purse, or e-cash products for small consumer and perhaps small business transactions.

By 2002 to 2005, however, digital money will become a large component of the payment process in Europe. This trend will be driven by Electronic Business, particularly Internet commerce and retailing.

Table 1. Worldwide use of digital money products (INPUT)

Digital Money Use by Banks	1997 (%)	2002 (%)
Money transfer, Swift Item	75	95
Bill payments: PC, Videotext, ATM Entry	70	90
Stored-Value Cards	30	70
Internet	<5	90

The reasons to move to digital money are relatively simple:

- Physical money is expensive. It costs US banks more than $60 billion a year just to handle cash. This includes an entire range of processes such as authenticating, counting, sorting and storing. It probably costs non-banking institutions in the United States at least the equivalent of another $60 billion a year to handle these cash systems.
- Cash is becoming increasingly easy to counterfeit. One of the main impacts of digital copying technology has been the facility with which currency and other financial instruments can be copied. There is a continuous war between counterfeiters and governments.
- Cash is the prime currency for one of the largest industries in the world: illegal drugs and other forms of crime. Without physical cash, many of these criminal organizations would lose their ability to function financially.

A key issue for banks is cost. Banks expect that digital money and online banking using the Internet, kiosks and smart ATMs will significantly reduce their branch expenses. On the other hand, they expect this new technology to add operational costs—in particular for data communications and networks. They also expect additional costs for marketing and support. There are additional costs attached to the security. These costs are typically not well planned.

As banks have moved into the areas of digital money and electronic bill presentment and payment, as well as into other areas, they have found that they also have to provide technology support. In fact, banks that have introduced online Internet banking services have found their support requirements changed dramatically. Essentially, if you are a client of an online, electronic banking, digital money product or service from a bank and, for whatever reason, it does not work, the first institution you are likely to call is the bank. The problem, however, could be with the software, the hardware or the network. Thus, banks have found themselves dragged unwillingly and unknowingly into the technology support environment. In many cases, they have sought to outsource portions of this support activity to technology companies. But it is very difficult to separate the technology component from the banking component in the electronic banking world. This problem is endemic to all forms of Electronic Business and is exacerbated by the fact that electronic support or electronic product support is almost invariably underplanned, underfunded and underappreciated. Yet its absence is what is most likely to cause customers to stop using a product or service or indeed to transfer to another institution or organization.

Banks of the Future

Banks will strive to transfer customer transactions from tellers to telephones, from telephones to the Internet and from the Internet to various forms of electronic banking that range from "micro-coin" transactions to large, highly profitable, investment-oriented transactions.

Banks will be in serious competition with each other and with nonbanks to offer retail and corporate customers innovative, more convenient products at acceptable cost levels. Stored-value "smart cards" that can be "loaded" with monetary value through a bank ATM, a home PC or hand-held terminal promise customers unprecedented convenience at low cost per transaction. So electronic banking will involve the replenishment of real and virtual smart cards.

As a result of digital money and electronic banking, "virtual banks" will begin to replace "actual" banks in the '00s. Virtual banks exist primarily in cyberspace and use telephone, Internet, wireless or cable connections to reach customers. They have no physical "bricks-

and-mortar" buildings, which are both expensive and unsuitable for the "anywhere, anytime" orientation of electronic banking.

These virtual banks will operate across geographic boundaries. After all, credit card companies do so today. Extension of the credit card concept will make electronic banking for corporations almost ubiquitous. These companies in electronic space will act as bill consolidators. Thus, businesses and individuals will not have to make multiple payments. As with credit cards, they will make a very few payments to cover multiple transactions.

Virtual credit cards will be key. The function of a credit card is not to identify the purchaser, but to identify the payer—usually but not always a bank. Provided the identity of the buyer is authenticated, the concern of the vendor/merchant is whether or not sufficient funds are or will be available to cover the transaction. This information must be obtained from the payer (bank).

Provided there is some form of identification of the payer available so that the merchant can ascertain the availability of funds, there is no need for the merchant to know either the identity of the buyer or the payer. That is the "value" of cash and of stored-value systems. They are anonymous. This makes them the preferred form of instrument for criminal activities.

Bio-identification processes (retina scanning, fingerprinting, etc.) will provide for "private" individual identification for personal and business purposes. Personal readers (bar code or other identifiers) will allow individuals to scan tickets on goods. Attached to communications devices (portable IAD), this will enable an individual to go to a shop, literally pick up an item and walk out with it without having to go through a check-out. Self-service will come to retail.

This is not so much electronic banking as it is electronic payment. This represents the 21st century trend toward process and away from industry.

The handling of payroll is another aspect of banking that will be particularly important. Many of you reading this book will have this day earned money through employment. When will you receive the money you have earned? Perhaps your initial response is "I will receive it on payday." But do you really receive that money? No. Your employer deposits money in a bank. However, does it really deposit money? No. An entry is merely made on a value sheet—a balance sheet. This entry says that you have the right to withdraw some value whenever you wish to from a particular account, in order to employ the value for whatever purpose you choose, whether it be to pay your rent, your automobile expenses, for movies, sports, etc. You actually do not get use of the money you earned today for a considerable period of time. What happens to that value or money while you're waiting?

The answer, from your perspective, is nothing. But from the corporation's viewpoint and from the bank's viewpoint, that money is available to be used.

Can the government collect taxes on it? Only when you are "paid," or shortly thereafter. Significant portions of national debts would disappear if governments could accelerate payment of taxes to the time of earning, instead of the time of payment or use. Most people would like the use of their money as soon as possible; so a reduction in the time float between earning and receiving would be attractive. If governments combined that benefit with automatic and early payment of taxes, they might have a politically attractive scenario.

Global banks report that the proportion of their revenues due to digital money or electronic banking products will shift from less than 3 percent in 1996 to about 20 percent ($300 billion) by 2001. The value of the electronic banking products and services delivered by banks at that time will easily exceed $1 trillion.

The current control of the money supply as well as the accuracy of its reported activities will be seriously affected by electronic banking products. Some government banks do not count credit card transactions as part of their money supply reporting. As the Internet moves more and more transactions to credit card or other virtual exchanges, the virtual money supply will become a significant fraction of the official money supply figures. The counting techniques, the central bank control processes or both will be changing soon.

Breaking down Barriers

Banking, securities and insurance are all concerned with financial instruments and money. Already, the boundaries are becoming less distinct between the various organizations that form these industries. Indeed, they are also blurring between these and other industries such as retailing.

The Internet will amplify and accelerate the breaking down of the boundaries of these organizations and also the blurring with other processes and industries.

For example, what is a bank? It is an organization that stores value. It is a sophisticated bartering mechanism. It allows individuals and organizations to develop value through their work efforts, value that can be stored until it is used to acquire a product or service from someone else.

Will Electronic Business allow other organizations to store value? Yes. Organizations, particularly large ones, could store the value generated by their employees' work efforts and sales, pay substantial interest on these values and negotiate with other organizations for special deals in order to transfer that value from one organization to another.

At some point, this value would have to be translated into money, at least for the foreseeable future. But just as there is a movement toward establishing super currencies such as the euro and (de facto) the dollar, there is also a countervailing tendency to set up smaller currencies.

Banks were extremely frightened, and many still are, of the power that Microsoft could exert in the banking business. Instead of having dollar bills, we could well have Bill's dollars!

But most major organizations are very capable of functioning as banks. The software systems, services and support are all available for them to do so. Many large companies such as BMW, GE and General Motors, have set up the equivalents of banks, although they do not act as retail banks yet. Credit unions established within companies of all kinds have some characteristics of banks. Regulation prohibits them from fully competing in the banking market.

Conversely, banks will have to reexamine their roles. One question facing banks that is not necessarily related to banking is to what extent they can and should use the new environment to develop and offer new products. Should they use their role as trusted agents to move themselves into other areas of the electronic marketplace? For example, should they become portals? There is a major debate in the United States over the direction of banks in this regard. Some of the newer, electronic banks are rapidly moving to offer enhanced services to their customers. Probably the most important facet of any individual's or organization's "life" is finance. If so, then an argument can be made that the financial service institution should become the core of the customer's future activities in any number of areas, particularly areas such as investment and insurance. Banks also see a rapid movement of other types of companies into their banking domain.

SOHO Electronic Banking

Consumer and small business electronic banking will become a major battleground and source of change in banking.

- SOHO (small office/home office) electronic banking payments will generate large savings in processing and operations.
- Additional products and services will be sold to bank customers who generate electronic banking payments.
- Bank and non-bank competitors will aggressively attempt to capture business with such services.
- SOHO electronic banking bill payments on the Internet will reach a value of $1 trillion in the first decade of the 21st century.

The potential benefits of Internet financial transactions and depositors' strong demands for electronic transaction services will drive electronic banking. Payments through SOHO electronic banking (which may be initiated from a home, hotel, car or small business location) will be a source of savings as well as a vehicle to contact customers and sell services, including maintenance or management of the account.

- A payment item from a SOHO electronic banking system will save $1.00 to $1.50 in processing costs if the payment can be delivered to the payee electronically. Payees will almost universally accept electronic payment by 2010.
- Savings to banks from average depositors (20 to 30 payments per month) will amount to hundreds of euros per year per account. These savings will have to be shared with depositors though a reduction in fees. Note that this will reduce bank revenues.

Payers will benefit through savings in postage and time (delivery), as well as through a reduction in bank costs and fees. Banks and merchants will realize savings in billing, payment and receiving when a payment is made online on the Internet. These systems will encourage switching of accounts. A bank depositor of Bank A might make payments online on the Internet through a service provided by an IT vendor and Bank B. Initially, Bank A will lose only a few payment items for this depositor, which wouldn't be of concern. But eventually, Bank A might lose all payment items for the depositor and eventually the deposit and other accounts to the IT vendor and Bank B.

Vendors serving the banking industry and some banks will use Internet banking payment systems to contact consumers and persuade them to transfer their checking and other accounts. SOHO electronic banking will enable non-banks to sell banking services directly to bank customers and possibly to take over management of the account for the individual business. Vendors will offer the account holder the opportunity to move the banking relationship to another bank in order to take advantage of lower fees or higher rates for savings or investment. Or IT vendors will manage the account(s) and select banks or non-banks that supply the best terms or features for any service or product needed.

Control will shift from banks to non-banks in many cases, particularly where banking services are offered "free" as incentives for other services.

The most common source of payment for SOHO electronic banking payment will continue to be credit card accounts. Some vendors will provide a credit card solution as well as solutions that involve the use of debit cards or cash accounts. Secure credit card processing capability will be embedded in Web servers. Credit card organizations provide standard solutions that will be a key consideration in their continued success.

There will be a huge battle to control the front end of the process at the customer's location. The battle will be between vendors (banks, services companies and others) and approaches.

In some approaches, the primary software and data for the customer will reside on the customer's system. It will often be independent of banks or payment processing vendors. Intuit, for example, already has 15 million customers for its PFS product. By 2010, this could reach 100 million. It also provides equivalent software products for small businesses, as do many other vendors.

This customer software will be provided virtually free. It will be much easier to use in 2005 than it is today. It will incorporate voice processing, agents and graphical processes to guide customers.

Since the software and data will be independent, they can be used by vendors to attack banking relationships and to enable customers to switch services easily. These services will be for banking, investment, insurance and other activities. The software will enable integration of financial products in ways that are simply not possible in 1998. Customers will be able to construct self-insurance programs, for example, in collaboration with other individuals and/or organizations.

In other approaches, the software will reside on a Web server together with the customer's data. The customer will use voice, key entry, interactive TV or other methods to initiate payments, with-drawals, investments, etc. These systems will be less flexible than customer site systems, but much easier to establish and use. This will be their main attraction.

Electronic cash and financial management for SOHOs will be a core part of the services provided. Services that have only been avail-able to large institutions and private banking (i.e. wealthy) individu-als will become available to all. If banks do not provide them, others will.

As "trusted agents," banks will provide electronic escrow services of all kinds.

Financial transactions for purchases initiated off the Internet will also be handled on the Internet. A buyer might want to think over a purchase before contacting the seller on the Internet to finalize it. A buyer might also want to use the Internet to interact with the seller and negotiate additional terms before buying.

Insurance

Three areas of electronic insurance are being accelerated. The first is the sales processes through agents and brokers; the second is claims handling, particularly in areas such as automobile and health care insurance; and the third is payment systems, particularly to and from individuals as opposed to corporations. Note that most of these serv-ices are targeted really at enhanced Internet-based insurance activi-ties, as opposed to new insurance methods that we will discuss in the following chapter.

These trends have already begun with the introduction of demand-based sales through devices such as kiosks, ATMs or home comput-ers. A good example of a demand-based insurance product is the flight insurance that is on sale at many airports. Today, flight insur-ance is obtainable as part of other transactions such as the credit card payment of the flight ticket. Another example is the insurance on rental cars. Such demand-based insurance products are extremely profitable for insurance companies, particularly when the use of elec-tronic systems makes the costs of fulfillment relatively low. Its very profitability has led credit card, travel and other institutions to form their own insurance companies specifically devoted to such narrow demand-driven product segments.

The trend toward direct interface and sales of insurance products is also becoming evident. Most importantly, the products and services used by customers in the insurance environment are acquiring a much higher degree of volatility. Historically, few people and companies changed their insurance on a continuing basis. In the next several years, the insurance industry will become increasingly demand-based. The levels of insurance taken by individuals and organizations will fluctuate dramatically over time as opposed to the current situation in which insurance products tend to be very static.

Travel

A healthy global economy and increased competition are driving the demand for business travel. Increased consumer wealth is driving tourism. But an excess of demand over supply has led to significant price increases in the last several years that will continue over the next three to five years at least. These are also showing an annual increase in the range of 3 percent to 6 percent.

As a reaction to these increases, business users in particular are trying to control costs and gain control over the process. Therefore, there are two major issues facing the travel industry. The first is the need at least to contain and preferably to cut expenses. And the second is to deploy new technology to meet the challenge to the industry posed by the Internet.

The Internet is the supercharger of the tandem advances in telecommunications and transportation that have taken place in the 20th century. Telecommunications changes have led to more travel, and ease of transportation has led to more need for telecommunications.

In the future, Internet-based services may well replace the need for business travel. Or at least they will reduce the potential growth.

In the consumer or tourism area, substitution does not look likely for a considerable time. What is more likely to occur in the consumer area in the near future is the development of technology-specific travel destinations, an early example of which is Epcot.

The fundamental challenge in the near future in the use of the Internet, particularly in the travel industry, is disintermediation of the costly process related to travel agents.

Airlines and hotels in particular see the Internet as an opportunity to reduce costs through commissions they have traditionally paid on the ticketing process. Airlines used to pay on reservations, but primarily now pay on tickets. Through the process of electronic ticketing, they are attacking the whole distribution channel. As a consequence, the agent is shifting from a provider of information and fulfillment service to a travel manager. Agents have to add value, not just process tickets.

Corporate buyers want to gain control over the process, not just simply to control costs. They want to ensure adherence to their policies as opposed to Web and e-mail–induced anarchy. Business travel is an area in which costs can quickly get out of hand unless controlled carefully. As a consequence of these changes, large organizations in particular are establishing their own internal travel management functions. Although these were initially travel agents, they are becoming travel managers. As large travel agents look to become travel managers rather than travel agents, they strive to outsource the travel management function for organizations.

In 1997, 90 percent of all airline bookings went through travel agents, the airlines' most expensive distribution channel. Avoiding this channel is not easy, for airlines have developed overly complex pricing structures that make it very difficult for individuals to purchase anything but the simplest plans via the Web. To make a significant impact on the amount of Web bookings, airlines will have to develop much simpler and more understandable pricing structures.

A discriminatory pricing structure is the reason why a travel agency can offer prices lower than even those directly obtainable electronically from airlines, hotels and car rental organizations through electronic travel agents such as Travelocity. These large agents are acting as aggregators. As buyers become more empowered over the next several years, this pricing structure will be attacked by individuals.

In all areas of travel—airlines, hotels and car rental—there is a trend toward providing the basic service at low or nonexistent profit margins, then charge for incremental services. Thus, hotels compensate for cheaper basic service by charging more for telephone calls, valet services, in-room meals and the like. Airlines charge for movie headsets, games or drinks. Car rental companies charge for navigation systems, telephone systems, insurance, petrol, etc.

On the one hand, technology and competitive shopping are being used to reduce the price of the fundamental product or service used. On the other, however, there is almost no information or management provided to the individual or the business for these ancillary charges, which are where companies make their profit.

Travel industry participants have well learned the old Gillette slogan, "Give away the razor, and charge for the blades."

As we move into the next century, the new e-travel environment will provide more transparency and return more power to the individual and the organizational buyer.

Travel agencies themselves face very significant problems. They must develop a new value proposition. For consumers, they must advance from being mere ticket providers to acting as travel advisors. For business, they must evolve into travel management firms such as Amex Travel Services. This will elevate the importance of supplying information for travel management and in such areas as tax law processes. They will also become marketing organizations. Traditionally, travel agencies have been demand-driven responsive organizations. In the future, they must become proactive organizations.

The transition from ticket supplier to travel manager will by necessity shift their focus from consumers to corporate clients. The key to success with the latter customer category will be reducing transaction costs and the indirect costs of managing extensive travel budgets.

The future for electronic travel agencies seems cloudy. It is possible that these organizations, which are really Internet-enhanced electronic travel agencies, will disappear in the 21st century. To be successful, they will have to provide significantly enhanced services.

Corporate travel managers are discouraging business users against using these services because of concerns over policy compliance, tax issues, etc.—not because of concern about the use of Internet technology services per se.

Recent articles exploring various costs of channels for tickets such as direct, physical travel agent and electronic travel agent, have shown that the electronic travel agents have no compelling price advantage to offer.

The fundamental challenge for travel vendors over the next few years will be to establish trusted relationships with their clients. These relationships are probably at an all-time low.

Travel companies, especially the airlines, have a very negative image with travelers, particularly frequent travelers. This has resulted, in the United States, in calls for legislative and regulatory support. There is a movement toward a traveler's bill of rights. Interestingly, aggregations of individuals besides corporations are also promoting the rights of the traveler. Much of this is organized around the need for better information.

Virtually all travel-providing organizations have vertical information systems that do not provide for sharing of information across boundaries. Yet the individual who is traveling (and it is always an individual who travels, never a business) is demanding that information cross the various internal operating boundaries of the respective travel providers. For example, the individual wants information provided seamlessly across different types of travel providers within his or her travel schedule.

In this regard, the United States is more advanced than Europe, in the airline world in particular. The ability to transfer flights from one airline to another is much more advanced in the United States. In Europe, such information barriers are used as mechanisms to protect carriers.

So, in the future, travel organizations will use the Internet and the Web to build relationships, segment their customers and add value appropriate to each customer segment. Characteristics that will be supported by information networks will be particularly related to option choice. They will allow travelers to define by voice their travel requirements and will produce for them an integrated package of services at various price levels. This aggregation of services in the 21st century will be the main battleground for travel providers.

In the electronic travel industry of the future, vendors will try to shift transaction costs onto travelers, through ticketing and/or cancellation charges for example. Agents will migrate into the business sector and offer high-value travel management services in order to survive. Buyers will demand these added-value services in order to justify their continued use of intermediaries. Individuals, or consumers, will go direct: This eventually will involve going direct to a flight, a hotel or a specific car, rather than to the intermediary organizations.

Retailing

In his book, *e-shock,* Michael De Kare-Silver defines what he calls "the ES Test," by which products and services should be tested against three criteria: product characteristics, familiarity and confidence, and consumer attributes. He uses these characteristics and a scorecard to weigh a variety of products and services in terms of their likelihood to be purchased through electronic shopping. This segmentation is extremely useful for retailers and others. It shows that e-shopping harbors high potential for repeat orders. Yet, a telephone plus a catalog is still as fast, if not faster, than e-shopping. And e-shopping really replaces catalog ordering.

Voice processing is an advance that will affect this industry significantly. The ability to sit in one's car and dictate to a system one's orders for goods, including quality and other characteristics, will prove very useful in the early 21st century. Initially, this interface will be with a human order-taker. Eventually, this will be with an AI system. Advertisements from the 1996–1997 time frame showing a housewife sitting at a computer in her kitchen ordering will seem anachronistic.

A good example of the trend toward replacing catalogs with electronic shopping is the fact that Sears—the grandfather of catalog retailers—now sells its Mastercraft tools over the Web.

The items bought or sold over the Internet fall into two categories: first, physical things that cannot be delivered over the Internet and second, intangibles that can be delivered over the Internet such as news, information, music, photographs, film, etc.

Seen this way, Amazon.com is not really a new type of book company. It is a new distribution channel. Books are still printed and CDs are still made. This will not be the cutting edge of the future. It is not necessarily cheaper or faster than other methods. Has it affected markets? Certainly it has not impacted airport purchases of books, for example. Has it increased the market? Probably so. On the other hand, has it increased readership? Probably not. Many people who have used Amazon.com and other electronic ordering systems simply have taller stacks of unread books. Many of these books may have been distributed to them by friends who wish to demonstrate the power of their electronic purchasing and distribution knowledge.

As far as Amazon.com is concerned, the question is whether or not it will be able to transfer its undoubted market leadership position into profit-making avenues. It is trying to build affinity with its customers as quickly as possible. Its brilliant concept of identifying the types of books that would appeal to its customers and communicating news and information to them goes a long way toward keeping its customers loyal.

Experiments in electronic retailing will not always succeed, for example, the failure of the shopping mall to translate into the electronic space. On the other hand, the information malls such as those offered by Yahoo and other portal companies work very well indeed. They provide a collection of information about very targeted subjects. They are gateways to communities of interest built around these subjects. Returning to Amazon.com for a moment, another significant advantage and benefit it offers its customers is the establishment of communities, that is, people with similar interests built around authors, subjects or other criteria.

The nature of the product or service being bought is also a factor in e-shopping. On one hand, there are big-ticket consumer products such as cars; on the other, there are minor consumer products such as newspapers. The key to success in the various strata of the emerging e-commerce world will be how the structures in each of these areas will change. In the automobile area, for example, dealers may well assume service functions and also possibly serve as showrooms. However, the "showing" of the product, the cars themselves, is undergoing great change. Increasingly, carmakers will exhibit their products in consumer traffic centers such as airports and shopping malls. Some manufacturers, notably Korean suppliers, have moved to setting up displays at parking lots, shows and other consumer traffic centers, completely eschewing the establishment of traditional dealer structures. This is direct selling by the manufacturer or distributor to the buyer. This lack of traditional dealership structures is already prevalent in fleet purchasing in some geographic areas. Therefore, we can expect that the "showing" of a product, which is a significant part of the selling or buying process, will also evolve. Two distinct forms of showing that will emerge are those that are directed at those people who have a defined need to buy today versus those who must be stimulated into a buying decision.

In the Internet or electronic retailing space, most services today are focused on those that already have a defined or highly developed need to make a purchase. This is excluding the auction activities that are perhaps much more geared to either collectors or casual purchasers. In this regard, we are differentiating between collector purchasing and consumer purchasing.

The Internet will have a great impact on specialty shopping such as Christmas shopping. This applies especially to fad items. Many people will choose to purchase these defined items over the Internet rather than fight the consumer traffic in shops or shopping malls, only to be unable to purchase the intended item or items.

Electronic shopping for gifts at holiday times is probably the most positive potential area for electronic retailing. Interestingly enough, in many cases price is not a prime consideration in a consumer's decision to use the Internet.

Such a significant portion of its business is built around the Christmas season that electronic shopping at that time will have a huge impact on the industry. This is because the overall margins of retailing are relatively low, at least in the United States. (They are somewhat higher at this point in Europe, but this will be impacted over the next several years by the incursion of organizations such as Wal-Mart.) Christmas and gift shopping is a very important contributor to the profits of these organizations. Thus, a shift of even a relatively small percentage of high-priced, high-profit items from physical channels to electronic channels will have a disproportional effect on the profits of these organizations.

Retail stores then get caught in a "double whammy." At Christmas, they may lose a significant portion of high-value, high-profit fad items sales because of the convenience of e-shopping. They also lose a proportion of the after-market sales channel, which consists of people seeking the best possible deal.

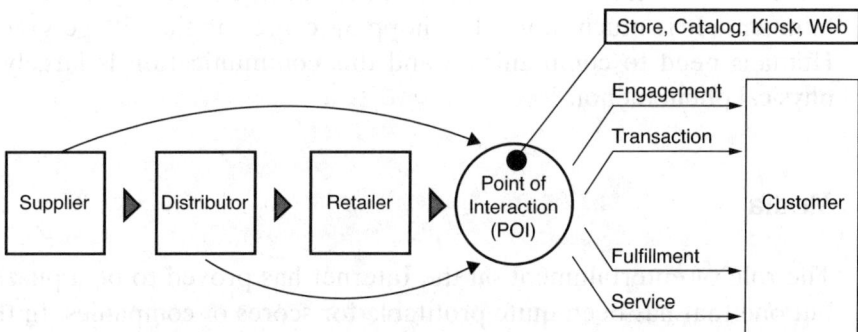

Figure 5. Using Electronic Business, suppliers and distributors can conduct transactions directly with their customers, bypassing retailers

Because retailers work on relatively thin margins, a small transfer of the orders of a few percentage points from physical shopping to electronic shopping should make a significant dent in their revenues and finally profits. These models would say that a reduction in revenues and margins at Christmas time would have to be accompanied by an increase in pricing on the remaining products, which would then prompt more people to shift to electronic buying, resulting in a vicious circle. But these analyses do not take into account that shopping may not be a zero-sum game. In other words, the transference of purchasing from physical to electronic medium may actually increase the total amount of products or services bought. As commented previously, Amazon.com has probably increased the numbers of books sold, though not the number of books read. It may well be in the manufacturer and retailer's interest to increase the amount of waste in our society. Thinly-stocked small refrigerators will eventually be replaced by overstocked dual refrigerators.

This scenario will translate over a short period of time into the value of properties such as shopping malls. We recently discussed with the owner of some of the largest shopping malls in the United States whether to hold or sell because, at today's values, these shopping malls have an unusually high price. Our conclusion was that if electronic shopping has such an impact on the impulse or gift-buying area at the high-profit periods, then the relative value of these physical malls would decline over the next five to 10 years and he should sell now.

However, as many commentators have advanced, shopping is not merely a means of survival, but it is also an experience, an entertainment and a social process. Technically speaking, this is not entertainment. It is much more the shopping center as the village green. Humans need to communicate and this communication is largely a physical phenomenon.

Media

The role of entertainment on the Internet has proved to be a puzzle, but one that has been quite profitable for scores of companies. In the mid-1990s, just as the World Wide Web was beginning to grow and solidify, a series of Internet soap operas caused a stir in the then-nascent Internet industry. Pundits were calling episodic content the next new Internet thing. Within a year, Internet soap operas had failed. They found small audiences, few continuing advertisers, and quickly receded to the background.

The object lesson is that many fads and forms will call out for attention in the entertainment domain. It is the goal of businesses to sort out the profitable aspects of content and entertainment that can take best advantage of all that the Internet has and will soon have to offer.

Publishers of books, magazines and other content have struggled too, because clear financial business models for new Internet initiatives have proved to be elusive, even to the largest players. The current favorite for most publications is a combination of advertising sales and shared e-commerce revenues. Some publications, such as the Dow Jones's online edition of *The Wall Street Journal*, have built-in audiences without much aversion to online content. But these are the exception.

Book publishing is another story, as the recent less-than-overwhelming introductions of e-book models demonstrated. Much corporate and organizational publishing has moved profitably online, with the Internet and Web easing distribution and timeliness issues, while reducing many other costs. The publishing of popular books for the general public, textbooks and other types of books is still a territory reminiscent of the Wild West: Pioneers are just as likely to have arrows in their chests as in their backs. No pure model is in the lead, nor will one dominant model necessarily emerge for all of book publishing. But there is a clear and growing trend for more and more information to be put on and accessed through the Internet.

Internet events are fast-paced, as the MP3 music data compression format issue illustrates. Recorded music companies, including some of the giants such as Bertelsmann, Sony and Universal, did not move quickly to arrive at standards for distributing online music. They hesitated for more than a year, despite the selection of the MP3 standard for audio and its pioneering use by independent music artists to distribute songs, even whole albums over the Internet.

Once Diamond Multimedia (now part of chipmaker S3) released its portable Rio player for MP3 music files, the record companies moved with more alacrity. But in the interim, millions had learned to use MP3 and may never turn back. The new industry standard will start out in 2000 behind the established MP3 standard. It is just these kinds of unpredictable situations that can foster the birth of new giants, even in industries in which the established players seem invincible. Internet time moves quickly and doesn't wait.

A big test of a similar nature looms on the horizon for the broadcasting industries (radio, TV, even movies and video rentals) and the test is broad: broadband or "pipes" wide enough to carry multimedia to billions of connected devices worldwide, and content in real time or as stored digital data. That broadband will "change everything" cannot be overemphasized. How it will do so is less clear. Network, even cable, video rental and satellite TV, will molt into a series of processes shaped by user demand instead of the current emphasis on producer schedules over constrained numbers of channels.

Billions will tune in to special events (news, sports, entertainment, etc.). Concurrently, smaller and smaller user segments will find and be found by their audiences. Subscriptions may take on a whole new meaning when Internet search facilities gain power and intelligence and more people are connected to the Internet. A producer (of a magazine, a play, a TV series, etc.) who can connect with n million people willing to subscribe for n cents or dollars presents a new model for financing entertainment and content in the age of the Internet.

When everything moves to the Internet and virtually every device can connect to it, new economies present themselves. Customization becomes a value added and a way to service ever larger and/or more well-defined audiences. The nature of entertainment companies, including advertising and other specialty industries, will be transformed by the new economic realities inherent in Internet-based interconnectivity.

Telecommunications

In a networked world, so much depends on the networks. Thus the future of telecommunications is a most critical part of the Electronic Business story.

The first task is to define a telco. This used to be a simple proposition; Each country had one telco, which was a regulated or government-owned institution.

Remnants of these organizations still exist. They include, in the United States, AT&T and "the baby Bells" that were created by the breakup of the AT&T long distance monopoly. There are also wireless providers: cellular, radio and other mobile organizations. Rounding out the telecommunications spectrum are cable providers from the television industry and satellite providers.

In the satellite area, there are GEOs (geosynchronous or high-Earth orbit), MEOs (middle-Earth orbit) and LEOs (low-Earth orbit satellite) companies. High-Earth orbit is the province of companies such as Hughes Electronic, and low-Earth orbit is the "potential" province of Microsoft's Bill Gates and his pals.

Satellite transmission, which many people had regarded as losing in the face of the proliferation of fiber, has recently been boosted by AOL's investment of up to $1 billion in Hughes Electronics' satellite capabilities.

Non-traditional telecommunications suppliers are moving into the market. These are utilities companies that include power, gas and water suppliers and transportation companies such as railroads.

There are community suppliers such as local governments. Many of these organizations are providing fiber-based transportation capabilities over their various rights of way.

This rush of organizations into the telecommunications infrastructure industry has resulted in an unusual combination of circumstances: There is overloaded bandwidth in high-traffic areas with low infrastructure provision. Elsewhere, there is a great deal of unused bandwidth available, particularly provided by fiber optics channels. This is so-called dark fiber. Significant costs are attached to "lighting up" this fiber, that is, making it useful.

Individuals and businesses will be presented with an increasing variety of choices in telecommunications. Today, these choices are presented mainly through direct marketing channels and advertising on TV and radio. This is using old technology to promote new. In the next several years, however, individuals and businesses will be able to buy their telecommunications services from telecommunications managers. This emergence of the Internet-enhanced business represents stage two of the Electronic Business environment.

New and old telcos will combine and join forces. In the United States, almost all the baby Bells have disappeared: They were too old, too slow and too backward. Most of the equivalent European organizations would also disappear if their governments did not protect them. Their business models are archaic. Their pricing is not cost-based and has been identified as such by the emergence of the new Internet organizations that provide voice under data.

It was not so many years ago that DUV (data under voice) best described data communications by the telcos. Now the age of VUD (voice under data) has arrived. All transmitted signals are becoming digital. It no longer matters whether we are sending a piece of data, a measurement, a piece of voice communication, a telephone call, a piece of video or a TV program. Furthermore, with the exception of voice, almost all communications in the past were one-way and not interactive—at least not in the sense of interactive being measured in seconds or fractions of a second, as opposed to days, weeks and months.

What is happening today to telecommunications companies because of the Internet is exactly what happened to computer companies 10 years ago because of the emergence of Unix and DOS from Microsoft. In both cases, external forces—today the Internet, yesterday the microcomputer and minicomputer operating environments—opened up hitherto rigid, closed systems. As always happens, the incumbents of those closed systems resist change. When change becomes inevitable, they are either forced out of business or forced to revise their businesses dramatically. In the computer industry, gross margins shrank from over 60 percent to about 30 percent in 18 months, according to John Acres at IBM. This is an incredible transformation. It is difficult to comprehend how companies could survive, let alone thrive, in that kind of dramatic change environment. Many computer systems companies failed to make the transition. Some survived because of government support but remain sick. They face eventual acquisition by or merger with stronger organizations.

The same thing will happen to the telcos over the next several years as a result of the Internet. It is not uncommon that companies in an industry experiencing such a transition are most successful just before the bottom falls out from under them. Just before the almost cataclysmic changes that affected IBM, Digital and other companies, some analysts were predicting that IBM would be a $100 billion company by 1990 and that it stood to dominate the computer industry worldwide. Profits and cash flows were at all-time highs.

The computer timesharing industry experienced a similar situation in the late 1970s/early 1980s before the advent of the personal computer. The most successful year for the timesharing industry was just before its calamitous drop in 1984. A similar situation is happening today with telephone companies in terms of profitability and cash flow.

The only difference today in many countries is that governments are willing to use taxpayer money to support structures that would otherwise disappear or be fundamentally changed.

The key area for the use of electronic telephony services is in customer care and billing (commonly known as CC&B). This application absorbs up to 30 percent of the IT budget in telecommunications companies, which are the most highly penetrated organizations in terms of their use of IT.

Customer care and billing is a very complex area. It is only a subset of a larger marketing segment for a telco because by definition CC&B addresses "customers" and not prospects. In fact, what has to happen in the Electronic Business world is that all customers must be treated as prospects.

What is happening in telco is symptomatic of what will happen in virtually all industries. Organizations are recognizing that the simple act of bill submission is now no longer a demand. If not handled properly, it is potentially a flag to the buyer that it is time to consider alternatives. It is also a significant opportunity to promote additional products and services. That is why for years additional marketing materials have been included in the physical bills that arrived in the mail.

In the future of electronic bill presentment, companies will make much more sophisticated use of this concept. Certainly those organizations that offer billing services may well provide them for free in exchange for the ability to access the customer from a marketing perspective to sell non-competing products and/or services.

Another huge area of opportunity related to telecommunications is the equivalent of the electronic yellow pages. These are the prototypical electronic information malls. The retail mall concept did not transfer easily to the Electronic Business world because shoppers have instant access to all e-retail possibilities. The Internet becomes, if you will, a single electronic mall. By contrast, the yellow pages concept transfers very easily. One could categorize the portal and search engine markets as being analogs of the yellow pages. When you want to find something, instead of going to the yellow pages, you go to a search engine. However, the physical yellow pages are still faster in access times than search engines for known information acquirements. And if you want to contact a company whose telephone number you do not know, it is still much faster to go to the directory inquiries than it is to try and get that information through the Internet, even if you are already online.

Yellow pages activities are one of the telephone companies' most profitable lines of business. Unfortunately for them, they have been extremely slow in moving this business into the Electronic Business world. This will hurt them in the long run, the long run here being three to five years, not 10 to 20 as they have been used to.

Need for Business R & D

So yes, there is a revolution. Yes, it is going to have an impact that is huge in the near future. But it is essential that buyers be very aggressive in protecting their interests by using only proven technologies, applications and services for core business processes and only experiment in areas that cannot do irrecoverable damage. We certainly believe that companies should experiment. In fact, organizations today need to have a business R&D budget as much as they have a technical or product R&D budget. They need to start experimenting with ways to work as opposed to just work to produce.

In the future, this activity will become a formal process with a knowledge base built around business experimentation. This has not really occurred since Industrial Revolution days. Over the next 50 years, new forms of business will emerge that have not been imagined. Others will change dramatically and some businesses will disappear because of Electronic Business.

6 Beyond Tomorrow

Because of the technology adoption curve, we used to be able to safely say that there is nothing that will affect a business significantly over the next five years that does not already exist in some form. But Electronic Business is changing so rapidly that this time period has probably been reduced in the United States to three years or less.

The key is to determine which of the developments happening today will be important for the future. For most companies, their course of action and impact on the market over the next two years is already in place. What they plan now will have its major impact in the period following 2001. This chapter provides some perspective on the conditions that will then prevail.

Electronic Business Perspectives for the Individual

Individuals will be able to enjoy an array of services through Electronic Business via the Internet. These services will include innovative options for active participation in entertainment, unique combinations of creative content and an "open" electronic education system attuned to the needs of the individual. In the paragraphs below, we'll take a closer look at some of these anticipated developments.

Power to the Individual

There are two types of Electronic Business: business-to-business and business-to-consumer. The latter category doesn't really do justice to individuals; it reduces them to being merely consumers. Each individual has characteristics other than just that of consumption. We study, communicate, love, worship, play, travel and rest as well as buy or consume. Many of the biggest opportunities in "Electronic Business" will not be about business at all.

Organizations of all kinds will use the Internet and other Electronic Business tools to build relationships with people in customers, suppliers, etc.; to segment customers; and to add value that is appropriate to each customer segment. (In this regard there are Electronic Business relationships to consider between and among government entities, individuals and business organizations.

Another area that can grow extremely rapidly is that of individual-to-individual or consumer-to-consumer. Sometimes individuals can be involved in dialog; at other times they may be involved, as in chat groups, in multi-path relationships.

The Internet-based individual-to-individual relationships can have business aspects. For example, individuals consult with each other. Initially this may be done at no cost, but eventually people will be bartering and selling products and services to each other. There is already a considerable volume of business being done on an individual-to-individual basis rather than through a formal organizational structure: Internet auctions are largely individual-to-individual transactions.

The phrase "business-to-consumer" betrays a certain arrogance on the part of business. The "business-to" part of the phrase is implicitly emphasized. In the future, even in the near future, the emphasis will be much more one of "individual-to-business" as we switch from a supply-driven to a demand-driven environment.

This is one of the most important changes embodied in the electronic world of the 21st century. Through the 20th century, democracies thrived and power moved from the hands of a relatively few individuals with good communications access to broad masses of the population in many countries. It is only in those countries with poor communications networks that we still have power resident in the hands of a few.

As this trend continues of power devolving from the few to the many, wherever they may be located, organizations of all kinds, including governments, will have to persuade individuals together or in groups to buy, use and/or follow their products and services.

All organizations consist of individuals. There is already today a blurring of the boundary between an individual's private life and his or her business life. At the extreme, this involves the process of teleworking when individuals work at home. We see the blurring of boundaries in areas such as the battles over frequent flier miles. When business people fly, they tend to accumulate miles in their frequent flier accounts. A number of organizations have tried to take these miles for their own use—after all, the individuals are being paid by the organization. Governments have tried to tax them as compensation. As affinity programs expand into areas other than travel, such as office products purchases or other purchases or contracts, the blurring becomes even more complex.

Another aspect of the blurring between business and individual or consumer is that many businesses today, particularly small business, use consumer-buying avenues. For example, Costco, a major wholesale price discount store in the United States, and companies like it are heavily used by small businesses, including restaurants.

Electronic Business will even further blur the boundaries among the traditional industry sector classifications and business process categories. Key industries and processes that will be affected by individual/personal Electronic Business activities include banking, investing, retailing, insurance, travel, communications, real estate and eventually manufacturing itself. By this, we mean individual manufacturing, what is traditionally known as "crafts."

Increased Fragmentation

Through the Internet, culture is amplified. We used the example of being able to watch a football team or an event of some kind anywhere around the world. However, this does not imply that there will be equal access to those events or sports matches. There will be increased fragmentation and discrimination among various levels of access to content of all kinds. Just as affinity programs were initially at one level and have since become fragmented into multiple levels, so will access to events become structured.

This structure will not simply be binary as to whether or not you qualify, nor will it be simply a question of payment, at least of a single payment. You could well choose to watch or participate in an event at multiple levels; i.e., you could watch an event at high-definition TV levels and pay a premium or you could watch the event at a slow-scan rate. We may even eventually have the capability to adjust viewing rate and payment rate at different levels during a particular event. If there was a special contestant or section of an event that a viewer wished to watch, the viewer could turn up the dial, the equivalent of turning up the volume, in order to improve the quality of resolution. It is this concept of buyer control over the level of access and payment that is most important.

The problem with today's models as described in the literature is that they are today's models. Future models will be of constructions we can only guess at. The above model has not only a pay/no-pay decision, but also a gradation of payments that are based on criteria that may include time, quality, level of participation, etc. A very rudimentary model of this can be seen in the way telephone charges have adjusted over the last five years with the opening up of competition. Whereas one could confidently say that in the past everybody paid the same rate for the same call, today that is not true. People and companies are part of any one of a number of different plans that may be related to coverage, timing, access capability, sign-up period, etc.

Participation

Instead of being passive, future buyers will participate in almost everything. They will certainly participate in entertainment. This will not simply be from the point of view of watching or being a voyeur. They will have the opportunity to actually participate in activities. They will be able to determine whether other members of the community are aware or unaware of their presence. This may be done either directly or through an avatar, an electronic surrogate.

Television stations today define their programming based on rating systems developed by organizations such as Arbitron and AC Nielsen. Future television program structure may well be defined by the direct voting of potential viewers. The issue will become whether some small set of experts determines a priori what the viewing community should watch versus the viewing community making its own decisions. This is the switching of the power from the producing activity to the buying activity.

TV stations and networks are not the producing functions; they are the distributing functions. The producing functions will increasingly participate directly through the Internet with their community. Already people such as cartoonists and some musicians are switching their whole distribution function to the Internet. This may work initially when the segments of the individual parts of the market are thinly populated on the Web. But once populations in particular areas explode to almost unmanageable proportions, there will develop a demand and a need for people and systems to assemble content and context in order to provide a viable product or service to a sufficiently large section of the community.

The public is not a homogenous whole. It is a collection, an increasingly fragmented or increasingly large collection of small communities or units that are in constant ebb and flow in terms of affinity, composition and size.

Personalization technologies will continue to gain acceptance until all online operations will incorporate some form of the technology. Given performance constraints on the Internet, it is unlikely that personalization will be true one-to-one interaction. Rather, users will be categorized into different "types," based upon common attributes. The user will see customized processes, site look-and-feel, and catalog views based upon the class of user, rather than a unique individual format. This method is more efficient and significantly less burdensome for Web servers than true one-to-one personalization.

Creative Content

To resume the discussion of the book business that we started by looking at Amazon.com, we should examine the creation of books as well as their distribution. Xerox Corporation, for example, provides new ways of producing "books" and publications. This change of production, probably better described as origination rather than production, will affect the whole concept of what we mean by "books." Furthermore, we will also integrate the different media of books, magazines, radio, TV and movies in unique combinations. Today, each one of these is a separate industry. Timelines, distribution channels, pricing, etc., are all different. Through the Internet and in the Electronic Business models, it becomes possible to take creative content and combine it with other forms of creative content in new ways. Thus, we may well see books produced that are accompanied by music. This will not be a movie; it will be a new medium.

One can argue that the poets of today, the Shakespeares of today, are the creators of our songs. Although some might sneer at this, we should remember that Shakespeare, Milton and others did not write their works for an "educated" public; they wrote for the masses. Today's popular rock songwriters also write for the masses. Songwriters and some artists are already taking to the medium of the Internet. Entertainers perceive the Internet as the medium of the masses, where masses equals audiences of tens or hundreds of millions versus tens or hundreds of thousands.

Thus, there will be new ways of producing creative content. We see an example of this in Linux, which is the software operating system created by communities in collaborative work across geographic boundaries.

So, what will we get in all this? For the individual, there will certainly be interactive entertainment including TV, movies, etc. Individuals and groups may well participate in these events or activities. Some activities will be transient and only retained if they are recorded; others will be works of art or entertainment capable of being repeated.

We will attach videoconferencing to all our communications activities as individuals or as groups. A video may or may not show us or an avatar.

Education

Education, which is either the largest or second-largest industry, depending on how it is defined, is probably the most susceptible to change and probably the least likely to. Established interests in the education community will make it very difficult to change education processes that have developed over centuries.

It has been said that if one took any profession from the 1890s and transferred it to the 1990s, the only one in which a practitioner could almost instantly function at 100 percent effectiveness would be teaching, provided the practitioner taught a subject, such as Latin, that has not changed. Teaching methods have remained the same. Certainly today, some technology is used to aid in the teaching process, but the process has not changed.

Electronic education is different. It allocates to each individual a unique set of materials and a pace at which those materials are learned and absorbed. Arguably, the main function of "teaching" is to help individuals work together in groups to solve problems; to help individuals overcome phobias and fears; and to help individuals and groups understand different points of view. This is much more like coaching than teaching. Another skill that is important to impart, not necessarily teach, is the ability to learn to change and adjust one's viewpoint depending on the presentation of new information.

Private institutions will likely be among the first to develop this kind of educational process. A preliminary example of the type of education that can be carried out is the Open University in the United Kingdom. This has had almost 2 million students. Recently, it opened a branch in the United States. The Open University functions primarily through a TV channel and accompanying material. The Open University invests up to $2 million in a specific course and recoups its costs by offering such a course via television. It has the facility to maintain the course as a particular area develops.

As this concept is transferred to the Internet with broadband capability, the opportunity for people of all types to gain access to first-class courses and indeed coaches will increase greatly. This promises to improve the quality of education and the dissemination of knowledge. Such an educational process is open to anyone regardless of age, sex, nationality, physical status, etc., with the exception of certain visually impaired people. The advantages of this kind of Internet-based learning versus television-based will be the introduction of time scales that are more attuned to the particular needs of individual students and the interactivity inherent in the Internet medium.

The interactive component in e-education is not simply interactions between "a teacher, counselor or coach" and the student, but also among and with peers. Quite often, the most effective way to gain an understanding of an issue, point or problem is not through teaching from "superior" sources, but through contact with peers. We've all participated as students in discussions with our peers in trying to solve certain problems. The Internet will provide this capability to all students involved in these open courses.

The boundary between research and learning, which has become quite rigid in most educational establishments, could well break down as individuals develop interests and pursue questions that arise from course material. If they feed this back into the educational system, the whole process can be improved through the participation of those who hitherto were merely regarded as receptors of the information and knowledge.

Community Managers

This returns us to the discussion of the interactive nature of the Internet and the communications that result. Communications itself is a business. At their lowest levels, chat sessions may become clubs that require joining fees. Such a club or group may be led by one or more individuals who are paid to act as facilitators. They may also be paid as owners, if they established the particular group.

Over the next few years, many of the portal companies being built around the concept of access and establishing communities of interest will transform into community managers. Access is likely to become a commodity. The subscription nature of a community built around a particular subject will be valuable. These communities are extensions of the bulletin board services that have existed for almost 10 years.

All that was required to set up a bulletin board service was a personal computer, some rudimentary software and access to the Internet.

The power of these communities, bulletin board services and chat sessions is recognized. We have a great deal of manipulation of such groups. This is particularly prevalent among groups that discuss stock markets and stock prices. Attempts by individuals and groups to hype stock prices lead to the promulgation of much false information.

Another area of chat is sex, which was discussed earlier. A major source of revenue for many of the companies in the information-distribution business, including AOL, is access to sex-related materials, chat groups, etc. This is unlikely to go away. It will probably increase in sophistication, size and scope. It has been said, "If you wish to see which technologies will be adopted in the future, go and look at what is happening in the pornography business; it is usually the first to adapt to technology." Particularly in the area of communications, there will be a great deal of innovative approaches to this subject.

Electronic Business Perspectives for Organizations

In previous paragraphs, we discussed the kinds of Electronic Business services individuals may obtain in the future over the Internet. In this section, we examine the services that businesses in particular will use.

The Electronic Commerce Landscape

It is the browser and the Web that give us access to virtually everything. It is the fulfillment of Tim Berners-Lee's vision of information available anywhere to anybody at any time. Because of this ease of access, a huge demand for aggregation and evaluation will be created. It is not unusual today to put a fairly narrow search criterion into a search engine and have several hundred thousand references identified. Even with further qualification of the search criteria, there will be an almost impossible number of "hits" for many inquiries.

The continuing explosion of information being added to the Web will exacerbate this situation. It is also being complicated by the fact that many organizations are resorting to subterfuge and sometimes almost illegal activity in order to link their Web site to popular phrases or search criteria. As an experiment, try searching for some well-known company names. You will be surprised to find that they do not appear in the first 10 or 20 hits that the search engines obtain. In fact, sometimes it can be quite difficult to get to that company's home page. Thus, we already see this battle for position in the Web space becoming, in some cases, quite dirty. This problem will worsen as use of the Internet spreads internationally.

One consequence will be the growth of the information mall. The information mall will be a collection of information about a particular subject area from a variety of sources that will develop a following as a trusted advisor in that particular space. Over the next five years, becoming these trusted advisors in the Web space will be one of the major opportunities created.

These virtual malls will collect information in and around a given subject. They will approach the subject from many different sides, including suppliers, buyers, users, government agencies, regulations, legal characteristics, taxation issues, cultural issues, geographic issues, etc. They will truly be sites for everything one would want to know about a particular subject.

An increasing number of organizations will use their electronic storefronts as factory gates. This does not necessarily mean dramatic price reductions at the factory gates. In fact, many organizations will keep their prices the same as obtainable through the distribution channel in order to retain profit margins and at the same time not upset their distribution channel. Some organizations will actually offer their products or services at a discount through traditional or physical channels in order to support electronic volume. Today the reverse is true: Companies look at their volume of electronic business as ancillary to their physical volume. In the next several years, crossover points will be reached where the electronic volume of business will outweigh in value their physical volume of business. That is the point at which pricing changes will occur.

The technology support for this Electronic Business change is here. There is a move toward server push and proactivity in the Internet Web space. Until now, it was often quite enough to put information and capabilities on your Web site, relying on search engines and other promotional activities to bring people to your site. This will not suffice in the future. Organizations will have to promote and push their services and products in a much more targeted, effective way.

One reason is that the search channels are getting jammed. As search channels lose their effectiveness and credibility as objective sources of reference, individuals and businesses will turn to other ways of getting information on what it is they wish to purchase or evaluate.

Server transaction management capabilities will support this push toward electronic commerce. Today, it is possible to buy almost complete electronic commerce systems from various vendors. These are still not industrial strength systems in the sense of their ability to process millions of transactions a day. But these "boxes" of electronic commerce capability will add features and functions very quickly over the next two years.

In the electronic space, security is paramount. Fortunately there is the ability to support almost any level of security, as discussed in Chapter 4.

Payment systems and electronic cash will become easily available. The whole banking process and banking system will start to change over the next two to three years. This will initially occur only in highly penetrated segments of our society. Some fairly sophisticated bartering systems and processes will emerge. After all, money or cash represents value that has been generated in one form or another and is owned by an individual or an organization. In time, traditional monetary systems will have a decreasing value.

Access to the Internet will become ubiquitous for most developed countries in the next several years. Initially, the Internet was accessible only by computers. In the early days, by dumb terminals working with minicomputers. Then the personal computer became the primary means of access. Over the next several years, this will gradually switch to handheld devices of all kinds, including pagers, palmtop computers, telephones, etc. This technology will enable many activities to be performed during what have hitherto been fairly inaccessible periods of our lives. For example, we will be able to carry out electronic commerce and electronic retailing transactions from cars, planes or trains. This will include making business decisions as well as individual purchasing decisions. Ergo, much of the processing power needed to make this happen will actually be resident in relatively small devices. This is a transition time; the real impact will be felt in the "far" future, i.e., in four to five years.

Much has been made of the possibilities of intelligent agents. Unfortunately, these intelligent agents in their early stages are pretty dumb. Certainly over the next 20 years, these agents will become more "intelligent," but it is doubtful whether they will pass the Turing test (i.e., of machine intelligence that can't be distinguished from human intelligence), except perhaps with very high-end computer systems.

Over the next several years, there will be a great deal of frustration with our information accessing capabilities. This will be due to the major changes in complexity, predictability and volume of the activities that information systems must support in order to handle Electronic Business.

Penetration of Electronic Commerce

Between 2001 and 2006, the value of e-commerce will almost treble across all goods and countries. In some areas, e-commerce trading will account for the majority of transactions by both number and value. In IT itself, for example, the proportion may exceed 90 percent by 2006. That is, more than 90 percent of IT transactions by value will be carried out via e-commerce.

As Figure 1 shows, by the year 2005, more than 40 percent of the 7 million business locations in the United States will be involved in some level of e-commerce, as compared with 5 percent today.

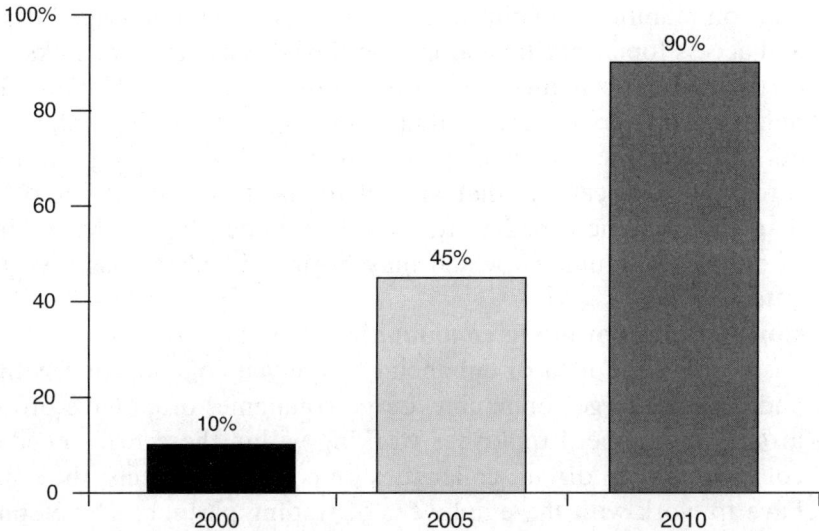

Figure 1. Use of e-commerce by US businesses, 2000–2010 (INPUT)

In the same time frame, large companies will use e-commerce to connect with more than 80 percent of their customer, partner and supplier base, compared with an average of less than 20 percent today.

Major changes in business processes will be necessary to support this. For large companies, the actual extension will be relatively easy at first as Internet technologies will allow them to extend their EDI networks to smaller companies via the Internet and Web technology. This will allow traditional trading partner networks to rapidly increase the size of their communities. Further on, however, these networks will have to be changed and the longer this is, the harder it will be to accomplish.

Electronic Trading Community Development

Community management and community development services will help companies move into larger trading communities. Community management falls into one of the following categories:

- **Web-based special interest groups**: These virtual communities are attracted to a central Web site by common interests that may be cultural, related to entertainment, career-oriented and so on. On-line communities communicate via chat rooms and message boards and access topic-specific features on the site such as news links, tutorials, advertisements and other tailored services. Community management services may entail attracting and keeping online users through intensive marketing efforts and developing site features and applications that appeal to the targeted viewer. Web marketing and new media firms will lead in developing these types of online communities, which may be individual- or business-oriented.
- **Collaborative corporate communities**: This type of community is based upon the idea of enhancing employee collaboration within and between large companies. Large companies often have offices around the globe. Employees working within these firms need to collaborate with distant colleagues on common projects; they may have to work with the employees of trading partners. Developing Intranet and groupware applications as well as common business processes can enhance cooperation between these separate groups. Community-building consists of working with companies to design, develop and implement scalable and extensible technology environments that provide members with the tools to converse, collaborate, grow and maintain internal communities.
- **Online trading communities**: This last type of community holds great interest. Companies will enter Internet trading communities

with their business partners. To effectively buy and sell from one another electronically, these communities will address all aspects of their interaction, including business processes, message formats, product databases, communications standards and so on.

Electronic Auctions

Online auctions (offers to sell) will gain momentum in the business-to-consumer market as companies such as eBay continue to expand product categories and services. Business-to-business auctions will have a relatively narrow focus, however. Businesses will use electronic auctions primarily to:

- Sell excess or out-of-season inventories
- Buy out-of-stock items from other businesses
- Run promotional giveaways to generate interest in a Web site or a specific product line

Online reverse auctions (offers to buy) of materials and services have not materialized yet to any significant extent. The Internet can lead to buyers' setting up virtual auctions where potential suppliers bid on contracts. This model could force suppliers to increasingly compete on price and could commoditize entire support industries. This has not happened yet because buyers are interested in establishing close relationships with suppliers in order to focus on issues other than just price, such as delivery times, supply chain integration and automatic replenishment contracts. But some examples exist and the practice will become more frequent as buyers become more familiar with it.

Electronic Business Changes the Marketing Industry

Electronic Business will affect virtually every type of marketing expenditure. It will also affect every aspect of marketing itself from the creation of material to its distribution.

By 2005, internal and external electronic marketing expenditures will potentially reach at least $100 billion per year. This is a staggering market. Often, the electronic component will be linked with the non-electronic component. For example, trade shows will have Web pages with videos of speeches and demonstrations that will also be carried on the Internet.

Expenditures
($ Billion)

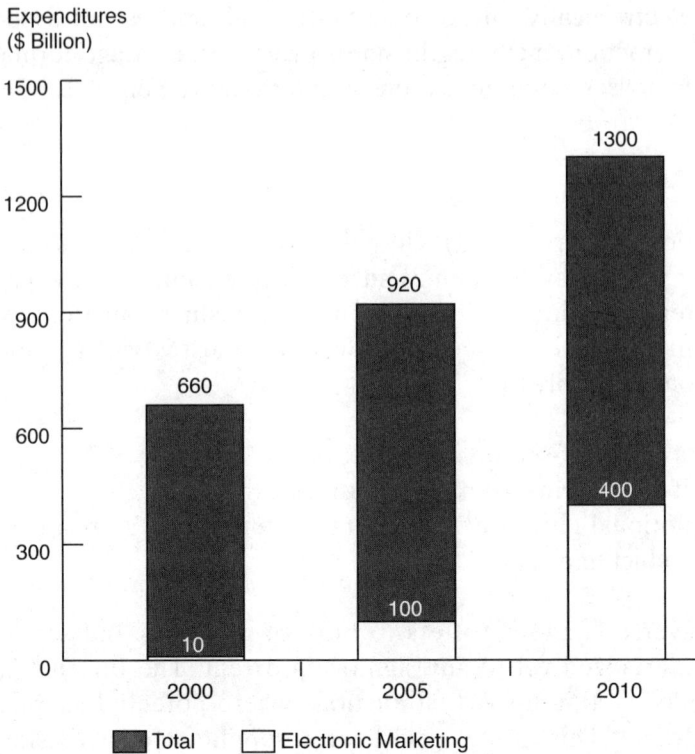

Figure 2. Worldwide external marketing expenditures, 2000–2010 (INPUT)

New Electronic Business Services

To achieve their Electronic Business targets, organizations of all kinds will need products and support from IT and communications industry vendors. What will they look for? What will become available?

Four Key Parameters

The new Electronic Business services will be built around the following key parameters:

- Security
- Scalability

- Value for money
- Payment methods

Security is a vital concern of any customer in buying IT and IT services for Electronic Business. Customers are looking for a security system that will provide protection against unauthorized and/or illegal access and theft and protection against loss of intellectual property associated with the products and services being sold, such as design information. They also seek protection against natural disasters such as hurricanes, storms, electrical storms, floods, etc., and protection against sabotage, terrorist activities, etc. In our view, electronic security will become one of the largest industries of the 21st century.

The United States has already experienced its first coordinated electronic attack on the communications infrastructure from an external country. Although this attack was fairly rudimentary and easily dealt with, it was a sign of what is to come. Attacks will be based on geographic, corporate, cultural and other indices. Companies and organizations in the Electronic Business world will seek a very high level of protection from their suppliers. This protection is not without cost and there is always a trade-off between security and ease of use.

The most important feature, after secure environments, will be scalable technology. With scalable technology, the software network and hardware involved in a particular system can adjust instantly, not just rapidly, to large swings in volume. It must be able to adjust for volumes that may vary by orders of magnitude over relatively short periods of time. This means the ability to move from a thousand transactions an hour to ten thousand transactions an hour and back again. It means the ability to adjust to rates of a hundred thousand hits per week on a Web site to millions of hits and back again. The system must not be so expensive that it impedes the cost-effectiveness of the service provided.

The penalties for being unable to provide service in periods of intense use are severe. The press is full of examples of Web sites that break down or cannot give service in periods of high activity. Once an organization loses a customer, it faces an incredibly expensive effort to get the customer back. Thus, scalability becomes a vital concern of any organization entering the Electronic Business space.

Scalability is not something that internal or indeed external IT organizations have traditionally dealt with. Except for a few historic cases (timesharing in the late 1970s and early 1980s, for example) almost all IT systems are built to handle predictable volume. They are built to handle seasonal or diurnal variations in volume, but these are generally quite predictable. In this new environment, organizations must deal with unpredictable levels of demand at relatively unpredictable times. This is because many of these systems will cover multiple time zones and changes in the environment may occur when people are awake at any particular point around the world. Global systems complicate the scalability issue.

Value for money is another concern for any customer of an IT vendor in this space. This differs from simple low cost. In fact, certain areas of the IT industry that were increasingly cost-driven are switching to being value-driven. Maintenance is one example. This is an extremely profitable area for most vendors, but it has been subject to tremendous cost and price pressure over the last few years. Because of Electronic Business and its requirements for 24-hour, seven-days-a-week processing, some buyers are now prepared to pay increased charges for business critical services with an almost 100-percent uptime guarantee level. In the Electronic Business world, one cannot afford infrastructure downtime. It is the equivalent of not being able to open one's shop in a shopping mall when customers are waiting, or worse.

Operational services will be paid for in many different ways. The first method will be the traditional long-term outsourcing contract. Initially, this may be relatively difficult to realize except in cases of co-resident outsourcing, in which the vendor and customer jointly develop the new business service.

Transaction processing activities will be another form of payment. This is the rebirth of what was traditionally known as the "service bureau." In this form of pricing, the customer pays by the transaction or by the value of the transactions.

Utility pricing is a method of paying for services that will become less common. With this method, the customer pays by the resources used for servers, networks, communications, support, software development, etc. This pricing system is unpredictable and this does not provide a good basis for an arrangement in a normal budgetary environment. But because of the variability and unpredictability of demand, utility pricing may well be an ancillary form of payment that is attached to one of the other two formats for long-run operation.

It will be quite common for many organizations to use external services providers instead of developing their own internal operations. Our forecasts are based on the same internal/external distribution of costs in the Electronic Business environment as in the old IT support model. But we expect that buyers will increasingly turn to external services of support for the Electronic Business world because of the huge and rapid potential variation in technology, demand, scalability, etc. The use of service providers is always greatest when there is the highest potential level of change. Buyers have a lot more flexibility in dealing with their providers on a service basis than they do with internal organization structures. As a result, our forecasts for IT services use in Electronic Business are probably low.

Internet Services Coming Your Way

Electronic Business will need a particularly strong service support structure that can deal with the changes in markets, regulation, technology and competition that will come with it.

**Internet Network
and Operational Services**

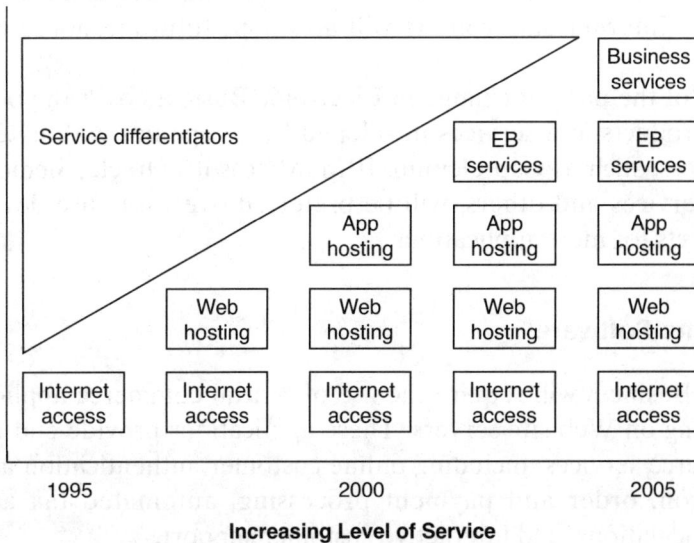

				Business services
Service differentiators			EB services	EB services
		App hosting	App hosting	App hosting
	Web hosting	Web hosting	Web hosting	Web hosting
Internet access	Internet access	Internet access	Internet access	Internet access

1995	2000	2005

Increasing Level of Service

Figure 3. The evolution of Internet services

In-House Versus Packaged Applications

Electronic Business applications development will rely far more on external products and services than IT application development with its large in-house staffs. Early adopters of Internet commerce developed their own unique applications internally for two main reasons:

- They believed their business rules were so complex that nothing "off the shelf" could come close to meeting their needs.
- True integrators had not come into the Internet space yet, mainly due to lack of sophisticated tools.

The market has addressed both these problems.

Electronic Business applications will be modular and more components will be bought instead of built. Companies, especially larger firms, will buy the basic functionality of an application and often continue to develop unique customer service functions such as account controls, order submission, order inquiry, price/availability queries, etc. Component-type Electronic Business applications will have software "hooks" that allow developers to easily integrate with a company's backend systems. The trend in outsourcing Web commerce functions to Internet data centers will also limit future in-house development.

Because of the pace of change in Electronic Business over the next 10 years, products and services developed by companies with extensive resources, such as Cap Gemini, IBM, Microsoft, Oracle, Siemens Business Services and others, will be preferred over in-house developed products for most applications.

Applications Software

Electronic Business will require the use of central commerce applications residing on Web site servers. These applications provide end-to-end commerce services, including online customer authentication and authorization, order and payment processing, automated tax and shipping calculations, and interactive customer support.

Applications also provide centralized procurement control to manage spending and acquisition, save time and resources, and negotiate price reductions from key suppliers.

Integrated decision support systems will be a key component. This will require integration of Web servers, applications servers and traditional databases to pull information from all three sources.

Market Size
($ Billion)

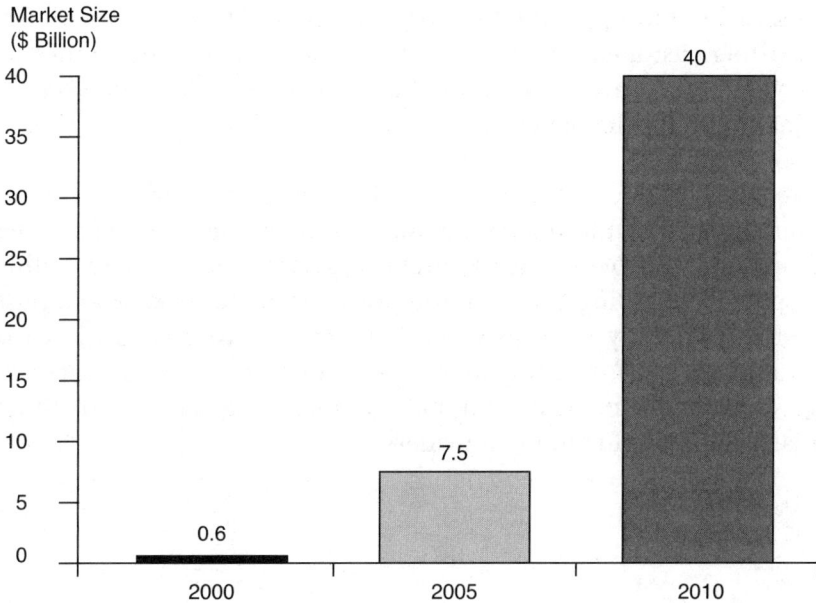

Figure 4. Worldwide Electronic Business applications software products market 2000–2010 (INPUT)

Rebirth of the Service Bureau

In the 2000s, a significant portion of Electronic Business will be conducted on hosted servers linked to the Internet. Many of the companies that host commerce Web sites are Internet Service Providers (ISPs) who have gradually expanded the scope of their services beyond Internet access and simple Web site hosting.

While Internet data center vendors are focusing on high-end business application hosting, the core business of Internet access services will continue to develop. These basic services will provide the funding for vendors to continually expand the scope and sophistication of their services as they target high-volume Internet commerce firms.

The Internet access market will be affected by:

• Market saturation
• Falling prices due to the commoditization of Internet access

- More limited opportunities to generate additional business from existing customers. Hosting services can be sold several times to a company; Internet access cannot be sold to the same level.
- Integration with telephony

As the volumes of users grow, Internet access will become as commonplace as the telephone and postal systems. The best-placed companies to provide such ubiquitous service are de facto utilities. Despite the ongoing fragmentation of national telephone and postal services (whereby state providers are opened to competition from private suppliers), this fragmentation is, on a global level, encouraging consolidation. Multinational organizations are controlling a greater number of national providers.

Market Size
($ Billion)

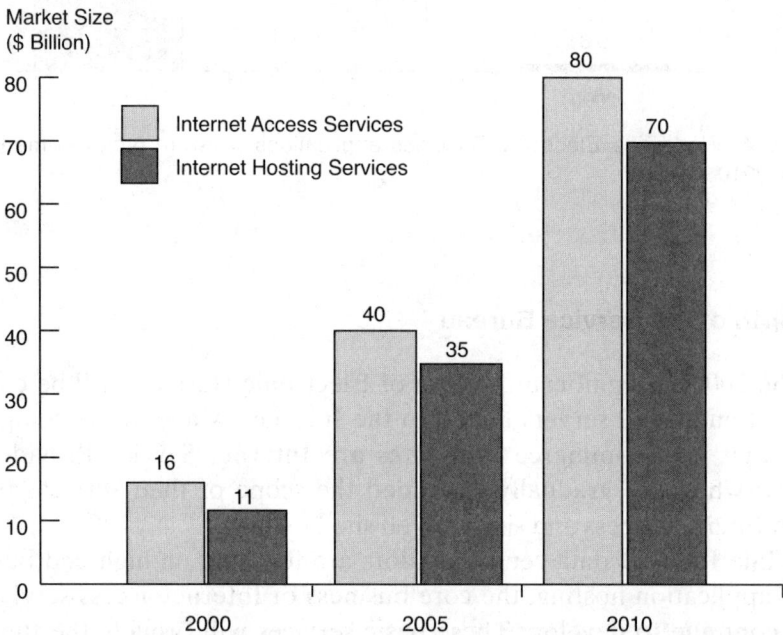

Figure 5. Worldwide Internet access services and hosting services markets 2000–2010 (INPUT)

The same will happen in the Internet access market, particularly in the residential sector. Customers will connect to, and be billed by, a regional or national company that will be part of a larger, multinational group.

Despite aggregation among large suppliers, the Internet access market will remain fragmented with many small ISPs in North

America and Europe. Many of them will serve small communities that are either geographic, cultural or some form of electronic community.

The other service that will fund ISP expansion into high-end commerce hosting is Web site hosting. In fact, most companies that have commerce-enabled sites are using the same ISP that was used for their initial Web sites. Figure 5 shows the market for Web and applications hosting services.

The market for all levels of hosting solutions, both simple sites and business-enabled sites, looks very solid. Like Electronic Business software applications, this industry is bolstered by ongoing shifts in computing technology and business processes.

Five trends are driving the increasing demand for Electronic Business operational services. First, there is the continued need to adopt Internet-based technology for networks and distributed information systems. Companies are shifting to Electronic Business processes and electronic commerce and want to focus on their core processes. Moreover, increasing numbers of users require commoditized support services. The increasing sophistication, and therefore expense, associated with scalable, full-feature Electronic Business applications running on the Internet is a tremendous challenge for many businesses. Finally, there is the rise of infomediaries and other electronic market places.

Strong growth of services, such as Internet Access and simple Web site hosting, and flush balance sheets will enable ISPs to quickly ramp up hosting facilities and expand the sophistication of their services. The Electronic Business operational services (Internet data center) industry is young, but has a strong outlook, as shown in Figure 6.

Currently, less than 5 percent of companies that sell on the Web use a hosted solution, but this will change significantly within five years.

As Internet commerce volumes reach critical mass, the level of service provided by Internet data centers (service bureaus) will be crucial to the success or failure of online sales. In particular, three basic areas of service will be important:

- **Facilities**: Security will be vital as companies sell an increasingly large percentage of their wares electronically. Internet data centers provide advanced "24 x 7" security features, such as fully secure cages and impact- and fire-resistant server vaults. Access is controlled and monitored by video camera surveillance and extensive alarm systems.

- **Network services**: Mission-critical Web sales sites must perform flawlessly even at the peak demand. Data centers provide network architectures that meet high volume demand through redundancy, distribution, and scalability. Excess capacity ensures that servers will respond to unanticipated spikes. Locating data centers close to public and private Internet exchange points minimizes traffic delays and optimizes performance.
- **Systems management and monitoring**: Complete monitoring of hardware, software, and site usage patterns enhances the security and stability of Internet commerce applications.

Market Size
($ Billion)

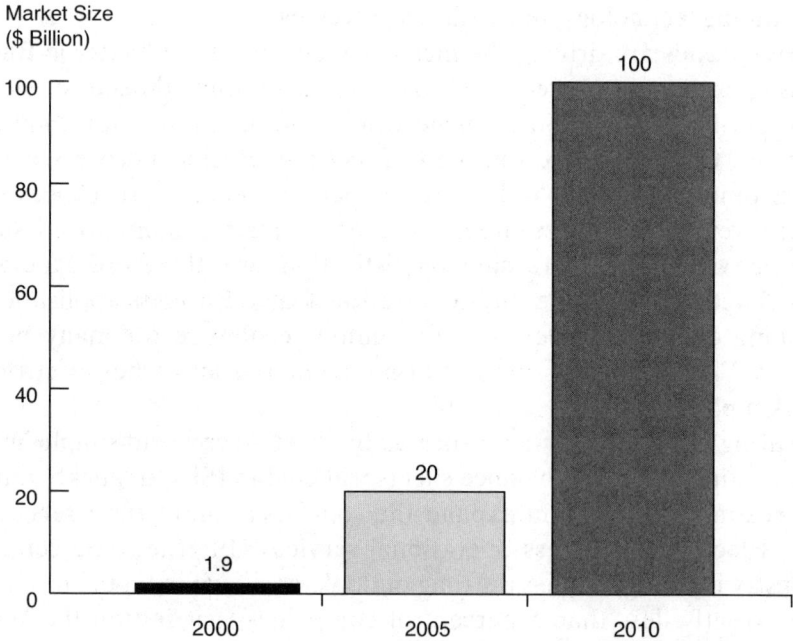

Figure 6. Worldwide Electronic Business operational services market, 2000–2010 (INPUT)

Emerging Application Service Providers

Application Service Providers (ASPs) are companies that host Internet-enabled backend and frontend applications such as e-mail, customer service programs and ERP applications. ASPs manage all of the software, hardware and network operations in their own data centers while customers access their systems remotely using standard

Internet and browser technologies. This business model is somewhat broader than hosting commerce Web sites because it extends core, mission-critical systems onto the Web.

Internet data center firms will play an important part in this emerging market. A typical arrangement might take the form of SAP, providing the Internet-enabled ERP package, Exodus running and managing the hardware and software, and a customer accessing the applications remotely via Web browsers. Internet data center firms will try to evolve into companies that provide a much broader set of outsourcing solutions for all critical portions of companies' IT infrastructure.

Consulting firms such as Corio will interact with and manage relationships with the customer, while relying on software from ERP vendors and network services from Internet data center companies.

Middle market companies with revenue between $100 million and $500 million are prime prospects for these services. These firms are big enough to need fairly sophisticated enterprise applications, but they don't want to dedicate extensive staff to IT functions. This middle market benefits from the trend in linking e-commerce applications such as electronic catalogs to traditional backend systems such as order entry, fulfillment and customer relationship management systems.

While the idea of Internet-based ASPs has generated excitement within the Internet data center industry, there are still significant issues to work out. ERP vendors must fully Web-enable their product lines in order for outsourced applications to run quickly and efficiently over distributed networks. Business processes and security priorities must be agreed upon by ASPs and users in order to iron out issues such as which employees within a user company have access to which portions of remote applications and databases.

Well-established IT outsourcing firms such as EDS, SBS and IBM Global Services participate in this market as well as the "hot" new companies such as Exodus.

Growth of Application Rental

Application rental is one step beyond application hosting. It requires the supplier to take a proactive, sales-driven approach, as opposed to the more passive technology-based approach of an infrastructure provider. Opportunities will become more numerous in the rental than in the hosting market, although unit sales will be lower. Buyers will use rental services to:

- Try out an application before purchasing the product and/or related hosting service
- Gain access to applications previously out of reach due to lack of budget or required infrastructure
- Provide access to non-critical applications for specialist staff otherwise too few in number to justify in-house implementation
- Cut costs, paying only for the functions they use and only for the time they use them
- Retain the option to change application without complex internal upgrades

Several hosting companies are entering this market in order to generate revenues and differentiate their services. Groupware and enterprise applications are early examples of rentable applications, and the variety will increase.

Rented applications are particularly well-suited to contract and freelance workers, grouped temporarily, who need to organize projects with disparate co-workers and who do not necessarily have permanent access to a corporate IT infrastructure.

Beyond rental, several major data center vendors, telcos, and outsourcers are working toward full utility services. The analogy of power utilities providing service via a national grid is a common one, and is becoming a more suitable model for the provision of IT as technology management increases in sophistication, remote hardware and software management becomes commonplace, and the Internet becomes the common medium for electronic services.

Trends and Issues

Early Adopters

The early adopters of sophisticated Web service offerings will be small and mid-size companies that have limited numbers of employees. These companies, which make up more than 80 percent of commercial sites on the Internet, have limited IT resources and will find it cost-effective to locate their servers in an Internet data center. The savings from volume discounts given to data centers are passed on to customers, and smaller firms will focus on the savings they gain in no longer having to pay high telecommunications charges because of their small size.

The availability of additional bandwidth as needed is another selling point of such a constellation. Smaller companies cannot afford to install high-volume lines for the few times during the year (such as Christmas) that their site receives peak volumes of visitors. As reliance upon Web sites for sales and customer service increases, larger companies will gradually move portions of their IT systems into hosted solutions. Eventually, only the largest, resource-richest companies will have all Web operations conducted in-house.

Internet-centric companies are also early adopters of these services. Companies such as Hotmail and Lycos, whose businesses are completely online, cannot afford to have problems with their site servers. They require extensive security, monitoring and redundancy built into their Web servers. This investment would be too large for a small company, especially an Internet startup that is experiencing rapid growth. Internet data centers have attracted many of these companies.

As data centers grow in sophistication and expertise they will expand into traditional industries such as retail, manufacturing and distribution.

Peering Agreements

The larger Internet data centers are rapidly evolving into exchange points where different networks pass their traffic to each other. Peering agreements provide fast, "clean" connections on the Internet between content holders, such as Exodus, and consumers who reside on other systems, such as AOL. These agreements reduce the number of connection points that a user must cross in order to reach the desired Web site. This increases the speed and efficiency of the connection and will be critical as more complex e-commerce transactions take place online. Connections will be established between competing data center firms and also among different facilities of the same data center company. Providing high-speed connections between the facilities of the same data center company will become more important as clients want to replicate their Web site server in multiple locations around the world.

Enhancing high-speed connectivity for Internet data centers will be required for next-generation Web applications. Fortunately, new telecommunication technologies such as Reflective Wave Multiplexing promise to increase bandwidth on existing lines by as much as a 40:1 ratio. As these technologies come into the market within the next two to three years, high-end vendors will be able to enter into high-quality service level agreements with each other. Consistent service levels between network vendors will reduce costs and greatly increase transmission and processing speeds to the point where high-bandwidth applications such as Voice Over the Internet, quality streaming video, and other multimedia applications become economically viable.

Electronic Applications Over the Next Five Years

This section summarizes application implementation over the next five years using five levels of complexity.

1. Web-based information systems

These are the easiest and most effective to implement on Intranets, they will include videos for corporate communications, contracts, partnering information to support the extended enterprise and financial information for internal and external use.

On the Internet, individuals will establish family pages and photo albums. Digital photography might be the "killer application" for the home. Individuals and small groups will establish periodicals and newspapers serving their families and associates.

Organizations will set up groups and teams that will work for customer benefit. They will have cross-function responsibilities that include marketing and support functions. These applications will be used to personalize services.

Performance data of all kinds will be made available inside and outside organizations; there will be nowhere to hide poor performance. This will make peer pressure more important.

Web-based applications will also serve investor interactions.

2. Web Integration applications

These applications will be bridge activities that will provide interaction with existing applications before native mode applications are developed for the Internet/Web environment.

These will also include the area of what is becoming known as "Business Intelligence." Data mining and analysis systems will be placed on the Intranet: that will use the vast amounts of data collected through the new electronic systems in order to adjust to changes in markets, competition and other environmental considerations as rapidly as possible. This will include new techniques such as 3D and MD (multi-dimensional) data analysis.

3. Workflow applications

These will include collaborative workflow connecting the various components of the supply chain internally within an organization. Parallel systems will provide collaborative workflow support externally to combine the efforts necessary to accomplish solutions for customers. This would include collaborative engineering and marketing for example. It will also include support for sales processes that include partners. Collaborative customer information systems are an example of the Electronic Business applications requirements.

These applications include demand chain automation that is the obverse of supply chain automation. Real-time monitoring of work activities will support inline adjustments to the work process.

4. Mission-critical applications

These will begin to achieve penetration in the next two years. Companies that have developed Internet applications for business will use their skills on a broader scale. Industries such as insurance have been slower than banking to respond, but respond they will, especially for activities such as claims processing. This will tie in to the health care industry and the increased emphasis on prevention and diagnostics that will occur in that industry.

Internet applications also support the trend in health care toward networks of providers that specialize in various disciplines rather than the hospital, "department store" approach.

For individuals, there will be growth in Internet-based entertainment, including the development of online communities, interactive broadcast services, reverse auctions and event/activity reservation systems.

5. High-bandwidth applications

Within five years, probably 30 percent of US homes and 10 percent of European homes will have very high-speed access. For businesses, these proportions will be doubled. This will open up applications

enablers such as desktop video and electronic video, document retrieval, or image retrieval.

Over the Internet, buyers will access broadcast "movies"; these
movies may be of any length from 10 minutes to 10 hours, or even
longer. Many of these entertainment services will be interactive, with
various levels of participation and pricing. An individual may be able
to buy the leading role opposite a "star."

Visualization will become a necessary component for business and
consumer choice. It will be used extensively for product demonstration and development. After visualization, participation will begin to
become an important characteristic, for example in race driving, role-
playing, conferencing, etc.

Networks will begin to become as reliable as electric power, and
applications that were not considered in the 20th century will become
common. Applications will include such highly sensitive areas as remote surgery. It will be possible to fly planes without pilots over the
Internet; it is a question of when rather than if.

There will be health monitoring for all of us over the Internet; initially on a daily sampling basis, but eventually using continuous
monitoring. This will not simply be for prevention of illness but also
for maximization of fitness, pleasure and effectiveness. Certainly we
will see this monitoring for high-risk categories in the 2000s.

Later On ...

Payment processes in general will improve. Payment integration
among trading groups will become possible. This will eliminate the
accounts receivable/accounts payable processes among them. A third
party could take over a clearing function among the trading group; all
balances will be available online.

This will remove the billing and collections process: When a payment is made at the end of the chain, its impact cascades along the
entire chain. A similar process is followed when a cost is incurred
anywhere along the chain. For consumers, it may be a bill consolida-
tor who handles the process. When a payment/deposit is made, a series of transactions is initiated automatically.

One area of interest to anyone and everyone both individual and
business is taxes. Taxes for Electronic Business can be collected
inline with the business activities. This potential acceleration of collection can make tax payments less onerous and improve the fiscal

performance of governments. It has the potential of changing our taxation process to a consumption-based and value-added process. Work and payroll can be included in this as we discuss later.

Sophisticated bartering systems will be employed; bartering typically does well in areas of high unemployment and low wealth. This will spread because of Electronic Business.

Organizations may become banks for their employees, contractors and partners; they will store the value of work and enable participants to trade with other organizations. The value of the work will be related to the value of the organization (easy to measure for a public company).

The financial reporting process is obsolete: The reporting systems and cycles we have today are too long. They hearken back to the days of the Industrial Revolution. By the time quarterly reports are issued, it is too late for investors. Auditing is too late as well. A year is too long; the function must go inline. With Electronic Business, earning can be calculated and announced continually.

Electronic education will eventually become a reality. Education will change to a more open system. Learning centers will teach people of all ages how to learn, where to go for information and how to use it. They will teach social interaction and team building. Their job will not be to teach facts; that will come from the electronic education system. They will teach students how to evaluate "facts," form opinions and express their ideas.

In telecommunications, each person will be able to have a single ID for phone (voice)/fax/video/data communications. A communications system will always be available to us.

Our friends, family members and colleagues will have ease in reaching us, no matter where we are. In dialing a person, we will simply use their name, not a series of numbers. Sophisticated call forwarding will allow the system to find us through our IAD, laptop or other device. "A phone will never ring in an empty room." If no one is identified as being active in the "room," the call will be automatically switched.

We will be able to speak in one language and hear in another. Speech recognition, particularly Speaker Independent (Voice) Recognition, will become available.

These kinds of value-added services will be offered over the Internet. Everything will be "intelligent." More important, almost everything of value will be connected. Drink-dispensing machines, for example, will be connected to the network for replenishment purposes and each machine will have the capability of placing a custom order.

No person will read a meter to input data into the system. All meters will be networked. This will enable energy, water and other scarce resources to be managed more carefully.

Will we need watches? Go into a room of 10 people, ask the time and the probability is that you will get 10 different answers. That is unnecessary with the technology today. A simple device can detect broadcast time signals so that we are all on the same time.

Questions about the rights of the individual versus the rights of society will remain. Battles over this and other political issues will be fought over the Internet, which will become intertwined with our political process. In the 2000s, we will see the first Internet-based elections.

Entertainment will become embodied in everything. We expect to hear music or get information when we are put on hold on a telephone call today. This process will spread. We can listen to music or watch an event while we are working. It will be difficult to distinguish work from entertainment in many cases.

In sports, the power of an athlete's name to attract customers to a product will increase exponentially. It is possible that sports heroes will command $1 billion a year including endorsements. We will be able to see the world through his or her eyes, to experience it and to participate in their activity. For this, we will pay a fee.

Electronic Business is based on teams that are multi-locational and multi-organizational. Players on these teams are in different companies or organizations. They are not confrontational but they are participatory. Organizations will be more open with their employees, contractors, suppliers and customers as well as advisors.

We all need governments and laws and to be part of communities. We are still physical and governed by physical laws; so the electronic and physical worlds have to coexist. The future offers the opportunity to blend the physical and the electronic in new ways that enhance our quality of life. We need countries, but not necessarily the ones we have now!

7 Myths and Imperatives of Electronic Business

Work is like water: It flows downhill to the lowest point of cost, consistent with a desired level of quality. Governments can build dams to hold back the flow, but work (like water) will find a way through, around or over eventually. Electronic Business provides the "hill" for the work to run down!

Organizations and communities must start now. Otherwise, they will be at a considerable competitive disadvantage and will miss the opportunity. Electronic Business will increase trade and business significantly. It will not be a "zero-sum game," as we said earlier. A rising tide will lift all boats—provided they float!

Electronic Business will provide access to worldwide markets for European companies, BUT they must be able to apply new technology and have excellent Electronic Business processes, particularly in sales, marketing and customer service. These have not been strong areas historically for Europeans who have traditionally looked down on service as being like servitude.

The Internet will change everybody and every business. Its impacts on people and businesses will make new Electronic Business and an electronic society possible. Yet some observers in Europe still denigrate it; they look for the negatives rather than the positives. There are certainly negatives associated with the Internet and the changes it brings. But that is true of almost all human inventions. They can be used for good or for evil. In this case, from an economic standpoint, the potential for good is enormous and avoidance will bring far more harm than use.

Beware the Myths

Separating hype from reality can often be difficult in this fast-moving world of Electronic Business. The media are full of predictions that the world economy is rushing headlong into complete submersion in the Internet where every individual and business is seamlessly con-

nected. It is wise to take a cold look at some of the biggest Electronic
Business myths:

Myth 1: It's Cheap

Despite many claims that Electronic Business, especially Internet
commerce, is a simple matter of installing a packaged electronic
catalog application and populating it with product data, true
Electronic Business will take a huge amount of time and resources.
Simple commerce-enabled Web sites can cost millions of dollars to
construct, while advanced high-volume sites can cost tens of millions.
Hardware costs are 10 percent and software costs just 15 percent of a
typical Electronic Business program. The largest amounts of money
go to integrating Electronic Business applications with backend legacy systems. Likewise, integrating outward to business partners systems is also expensive. Only by integrating with backend systems can
the full benefits of Electronic Business be realized and essential functions such as order status and product availability be incorporated
into commerce sites. Integration is the hidden cost that is never highlighted in vendors' pitches, but will prove to be a huge source of future income for systems integrators.

The amount of marketing and advertising required to attract customers to commerce-enabled Web sites is another source of hidden
cost. Companies have found that they must expend significant resources through traditional channels such as print media, television
and radio in order to raise awareness of their Web sites. Of course,
many of the current Web marketing campaigns are part of a mad
rush by Internet firms to establish a firm lead in their markets by
obtaining and locking in crucial early adopters. Even so, marketing
expenses for Electronic Business will be much higher than for traditional business.

Myth 2: It's Easy

Electronic Business seemed easy when it was viewed as a small,
somewhat esoteric portion of a company's operations. Now that
Electronic Business has gained a vital role in the future of many
companies, there is a realization that building an effective Web infrastructure is a large task. Early efforts at Internet commerce usually
involved a simple catalog on a typical "brochure-ware" company
Web site. However, as transaction volumes increase and the true

potential of Electronic Business becomes known, companies realize they will have to fully integrate their Electronic Business applications, produce truly scaleable solutions, and reformulate their business models.

The most difficult part of Electronic Business is not in the technology and integration area. The hard part comes when companies have to examine fundamental business assumptions such as customer and supplier relationships, distribution strategies, marketing philosophies, and other critical business processes. Ironing out these issues as Electronic Business technologies rapidly evolve is not an easy task, and companies are operating in uncharted territory.

Most of the failures in company Internet operations can be linked to several areas:

- **Poor design:** Companies often developed sites that would not scale and rapidly collapsed under surging volumes. There were also basic learning curve problems in designing Web sites, in areas such as download times, hard-to-find information and poor navigation rules. These problems are being resolved as industry gradually learns what works on a Web site and what does not work.
- **Lack of integration:** As mentioned previously, lack of integration with backend systems can cause severe headaches. Customers need to be able to perform self-service functions such as checking on order status and inventory. Also, sites that don't connect with backend systems mean that orders taken from the Web often have to be manually re-entered into order fulfillment systems. Luckily, most companies seem to have understood that integration is crucial to the success of an Electronic Business effort.
- **Poor communication:** Many sites do not have sufficient communication links with the customer. An effective commerce site must acknowledge receipt of an order (often via e-mail) and must be able to provide details of the order processing, such as order tracking number, shipping dates, etc. The order taken online must also be linked to the legacy order tracking system, so that if a customer calls inquiring about an order, a customer service representative can access information on that order by using the traditional tracking system.
- **No links to international sales:** One of the main lures of Internet sales is that a company's site can be accessed from any location. Therefore failure to service out-of-country visitors means a company does not receive the full benefits of Electronic Business.

Myth 3: Electronic Business Leads to Disintermediation

Electronic Business will allow producers to link directly to their end customers and in the process cut out conventional distributors, resellers and other middle-layer companies. In general, this has not yet happened. Many intermediaries exist for a very good reason—they add value to the customer purchasing process that the original producing company cannot or will not fulfill. Customer-oriented value-added services such as education, installation, repair, etc., are common functions of many intermediaries and are outside the core competencies of most producers. The Internet often acts as another tool for the intermediary, not the producer, to increase value to the end customer. Additionally, companies that are selling large volumes of product on the Internet, such as Cisco, often make the majority of sales to their existing intermediaries.

Electronic Business is creating a new breed of intermediaries on the Internet that bring together widely dispersed groups of buyers and sellers. These companies, called "infomediaries" or "market makers," are focused on specific vertical markets such as electronics, metals and chemicals. They add value to the process by acting as clearinghouses that collect information on what product suppliers wish to sell and match this information with interested buyers. Infomediaries collect consistent and comparable data on sellers' products and also provide tools that assist buyers in comparing and analyzing the sellers' offerings. Infomediaries typically make a profit by taking a share of the completed transaction.

The three basic types of infomediary include:

- **Lead Generator:** A company such as Autoweb.com that matches leads (car buyers) with suppliers (car dealers). The actual transaction is completed off-line, and profit can be made from a percentage of sales or can be a fee based upon each referral.
- **Market Exchange:** An online service with which an individual or company posts a request to buy or sell a certain product. Buyers search for products with certain criteria, and the system matches their requests with available products that meet their criteria. An example is NECX, which links electronic parts buyers with sellers.
- **Virtual Distributor:** Acts more like a traditional distributor. This service allows a purchaser to choose products from a large variety of vendors and add them to an online "shopping cart" and may offer a single payment mechanism to pay for all of the products. An example is Chemdex.com, which offers a convenient one-stop-

shopping Web site for purchasing a variety of research chemicals used by university scientists.

Finally, the fact remains that many producers can't afford to alienate their traditional intermediaries, yet. This leads to stopgap types of measures. For instance, most car companies will allow consumers to pick and choose a car from their Web site, but once a final choice has been made the site directs the consumer to the nearest dealership. This type of arrangement will be common for some time as industry groups struggle to define new relationships and business models evolve based upon increasingly sophisticated commerce technology.

Myth 4: Electronic Business Commoditizes Markets

The Internet increasingly enables customers to compare rival products. This leads to the theory, often called the "brokerage effect," that the Internet weakens traditional buyer-seller relationships by reducing the cost of searching and comparing products to the point where a customer can easily switch products based upon common criteria, most importantly the price of the product. The brokerage effect leads selling companies to increasingly compete based upon price and eventually commoditizes the market. Proponents of this theory point to agent technologies that will allow easy comparison of rival products and companies such as Priceline.com where consumers set the price they are willing to pay for a product and then let vendors compete to meet that price.

So far, the Internet has not led to widespread commoditization. For most consumers, there are many other factors that are just as, if not more, important than the final price of a product. Factors such as vendor reputation, quality, reliability, service guarantees and delivery are just as important to an Internet shopper as they are to the person shopping at the mall.

On the business-to-business side, theory would hold that large buying organizations would turn to Internet-based auctions where suppliers would bid on a specific contract and increasingly compete on price. In most cases, this has not yet happened. Some companies have found that Internet technologies actually strengthen their relationship with traditional business partners. For corporate purchasing, the use of Internet commerce applications has enabled supply chain integration and reductions in the number of suppliers. This has significantly strengthened the relationship between the surviving suppliers and the buyer. These suppliers are becoming true strategic partners that send and receive previously secret company inventory and production data via integrated systems. Of course, reductions in supplier base mean some companies lose their contracts, but these tend to be firms that do not have an Electronic Business strategy and are not equipped to support their customers electronically.

Thus, again the middle gets squeezed. Suppliers, particularly distributors, either have to become high-value-added partners or they will be eliminated. In the case of distributors, the buyer going directly to the sources will replace them.

Myth 5: Web Sales Must Cannibalize Traditional Content Sales

This argument has particular resonance in the publishing industry, where companies are struggling to place publications online and convince customers who are used to free Internet content to pay for them. This is especially true for newsmagazines and newspapers, with most experimenting to see if Web advertising makes up for potential losses in actual magazines sold over the counter. Today there is a wide range of approaches. Some publishers are putting all content online, others placing only selected articles in order to entice readers toward full subscriptions, and a few companies (such as *Wall Street Journal Interactive*) have successfully convinced readers to pay in order to see any content.

An interesting experiment conducted by Headland Digital Media, publisher of the *Rough Guide* series of travel books, is instructive. Headland decided to place all of its content online free of charge. Doubters claimed this was suicidal—readers would simply download and print content they were interested in and would therefore not purchase the books from local bookstores. In fact, the opposite happened. Web surfers viewed the Headland site, liked the content they saw, then proceeded to purchase the books at retail locations.

The lesson is that free content on a Web site does not necessarily damage sales through traditional channels. It depends upon the uniqueness of the content (an area in which newsmagazines often suffer) and the intended use of the information. In this case, consumers were going on vacation and therefore needed to take the travel-related information with them. Carrying around a neat, compact book made more sense than dealing with a large number of hard-to-manage printouts from a Web site. The free display and marketing of materials on the Web resulted in a boost to sales through traditional channels.

Electronic Business Imperatives

What must we do?

First we must plan: Consider "what-if" scenarios, track what is happening! And we must begin now.

This process must take place throughout an organization—not just in one research-oriented unit. This is perhaps the most important aspect of moving into Electronic Business. The whole organization is affected. Therefore, the whole organization must be involved.

Look at what Jack Welch has initiated at GE, arguably one of the best-managed companies in the world. According to *The Wall Street Journal* of June 22, 1999, he ordered his company to "Storm the Internet" in January 1999. Each of GE's 12 units has a team devoted to "destroyyourbusiness.com," examining how the Internet and Electronic Business can be used to attack their existing businesses and get them into new ones. This is called "eating your children before someone else does."

It's that serious.

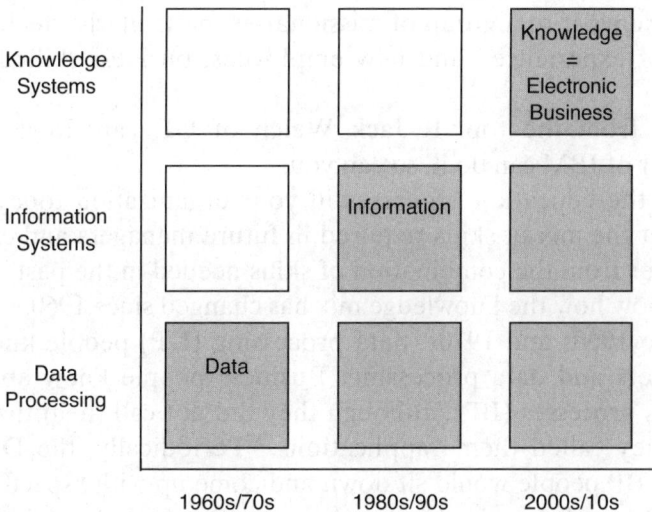

Figure 1. An IT organization's knowledge requirements, 1960s to 2010s

The changeover must be done in a controlled manner—not as some have done, without planning and almost without thinking. At the moment, customers and suppliers are willing to forgive mistakes as they go through the same transitions themselves. Indeed, they hope to learn from them. But this phase is quickly passing. Soon, the level of expectation will rise significantly. Thus, "Make mistakes now, not later" should be the motto.

From a technology viewpoint, getting started is fairly straightforward. Most companies have already taken some steps. Set up Web sites in each unit with good Internet access. That means perhaps with an ISP that is global, not local. The system must be scalable, secure and offer good service. The Web sites must be coordinated, particularly from a look-and-feel perspective. Set up internal Web sites as well as external. Force all your employees to use them. Use tricks, such as paying expenses immediately for expense reports submitted through the Web, to encourage Web site use. Allocate adequate resources to continuously update and maintain. Use tools and available external resources. Do not fall into the trap of trying to do it all yourself. Outsource wherever possible, so you get flexibility and exposure to the latest capabilities—but make sure that your business people are involved. Set up partnerships with Web service and product companies. Get them to help with the education and training process.

Do not leave this Electronic Business development to a sub-unit. Do not leave it to a group of missionaries, particularly "techies." Use teams of experienced and new employees, business-skilled and IT-skilled.

Lead from the top. If Jack Welch of GE can do it and Lou Gerstner of IBM can do it, so can you.

Start the education processes in your organization today. Recognize that the mix of skills required in future managers and executives will differ from the combination of skills needed in the past. Figures 1 and 2 show how the knowledge mix has changed since 1960.

In the 1960s and 1970s, data processing (DP) people knew about computers and data processing. Business people knew about their business processes (BP), although they did not call them that; if anything they called them "applications." Periodically, the DP people and the BP people would sit down and come up with "specifications." The DP people would develop systems they hoped would meet the specifications and quite often the systems did because the DP people were, by and large, smart and dedicated.

In the 1980s and 1990s, this scenario changed, at least in the more advanced organizations. Information systems began to be recognized as something more than glorified calculators with sets of accounting files. Concepts such as "Mission-Critical Systems" were lifted from the US Department of Defense and transferred to the commercial world, to be followed closely by "systems for competitive advantage" and other concepts. Businesspeople began to recognize that IT and IS could change processes as well as support them. So they started to look at business systems and the interaction of IT with their processes. Business schools such as those at Stanford and Harvard began to teach their students about the importance of IT.

Meanwhile, the IT professionals recognized that to be useful they needed to know a lot more about business processes. So they moved from just knowing about the data and its collection, processing, and dissemination to understanding how their technology could be used to handle information and provide business information systems to their users.

This progress can be tracked by looking at the consultants and IT services companies. In the 1970s the IT companies employed only computer people. By the end of the 1990s, they were employing large numbers of people whose basic skills were not in IT, although they knew how to use it. Some of the most aggressive recruiters at business schools became IT services companies.

At the same time, many of the traditional consulting companies recognized that their clients required that they become familiar with IT and how it could be applied. So they too moved into the Information and Business Systems environment.

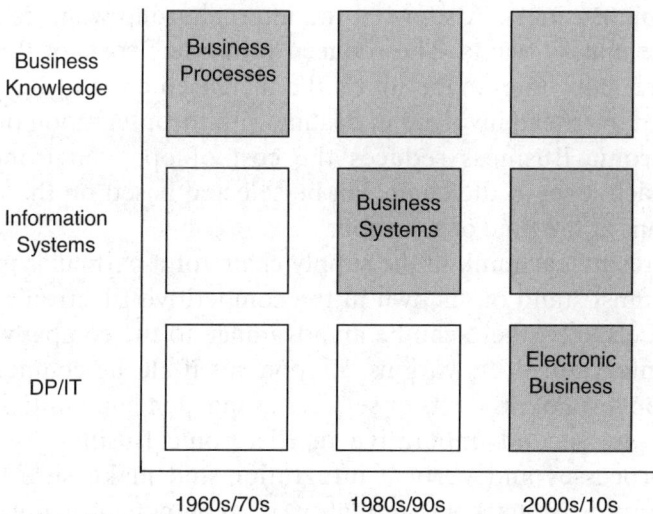

Figure 2. Business organization's knowledge requirements, 1960s to 2010s

Now we have moved on again. The gap between the IT and business people has virtually disappeared. Both must understand the knowledge and Electronic Business environment; these are now two sides of the same coin that cannot be looked at independently. Business people must understand technology and the IT community must understand the processes and how Electronic Business works.

Imperative 1: Eliminate the IT Planning Function

The IT planning function now has no useful purpose as an independent entity. Integrate it with the business planning function. Now! Some will say, "But there are so many technical things we have to worry about that should not have to be looked at in a business context." Really? It will be hard to find one that will not have some Electronic Business implication.

This reemphasizes the importance of IT and IS in the Electronic Business world.

Imperative 2: Break Apart Processes and Vertical Integration

In general, Electronic Business destroys vertical integration. It eliminates the advantages a company used to have from knowing what the next stage in the chain needed and from the reduced costs of sales and purchasing inherent in vertical integration.

The knowledge of what is needed is available easily today through Electronic Business. Also, what the internal group wants is often not what the market wants. The reduced processing costs of the internal chain are more than made up by the increased costs of lack of flexibility and overhead involved in dealing with the integration process.

Electronic Business reduces the cost of open environments in which each level in the chain can be selected based on the operating conditions at the time of selection.

As a result, each link in the supply chain for a particular product or service must stand on its own in the competitive Electronic Business world. Certainly, there can be an advantage to the company that acts as the integrator of its various components if all the components are in the top level of products or services as independent entities.

Thus the second imperative in Electronic Business is to break apart processes and vertical integration and make sure that each component is of marketable quality. If a component is not, then use

Electronic Business to make it so or get rid of it and buy a product or service that is.

Imperative 3: Don't Get Caught in the Middle

Invariably, the middle gets squeezed in Electronic Business. This is happening today in the personal computer business, where the packaging and distribution aspects of the industry are getting squeezed from the technology suppliers such as Microsoft and Intel on one hand and the solutions organizations, including in-house units, on the other.

The key in the Electronic Business age is to own either the customer or the basic technology; do not be merely the packager or the distributor unless you can add considerable value in the process or be able to use the process to generate something (usually knowledge) of high value. Portals are intermediaries that are giving away services in many cases in order to generate items of (supposedly) high value, namely "eyeballs."

Imperative 4: Advantage and Structure Are Transient

How long will this be of value? Probably not very long because switching technology and the ability to go directly to the source will improve.

Recognize that advantage and structure are transient. Many people hope that we will go through a period of sharp change and that then things will "return to normal." Sorry, but that will not happen. Change will be constant. It will come from unexpected sources and at unexpected times.

Channels will have to be constantly evaluated, for example. Today your channel sales, distribution and support strategy may be fine. Tomorrow or next year or the year after, it may be obsolete. Your basic operating structure is a transient state, so determine the optimistic and pessimistic scenarios (just two) of its survivability and stability. Then determine your actions.

Imperative 5: Operate on Your Customers' Time Scales

Many companies will have to rethink their time-related operating processes. French and other European companies could typically almost close down in August. Will they be able to do that in the world of Electronic Business? We think not. Many companies will have to go to some form of 24 x 7 x 52 operation; that is, 24 hours per day, seven days a week and 52 weeks a year. If you have customers around the world then you must operate on world time scales. This is particularly true for support services. That answering machine message that says, "We are open from 9 to 5 on Monday through Friday," will simply disappear, or the company that uses it will disappear.

Operate according to the local time scales of your customers and prospects—not your own.

Imperative 6: Track Your Competition

In Electronic Business, your competition will not always be who you think it is. In particular, new competition will come from companies that work with the "arms manufacturers": the IT communications, software, systems and services companies that are developing the tools for the Electronic Business world. Competitors will come from companies that work with new technologies; these may be brand-new, venture-financed companies as well as old. Recognize that there are billions of dollars available to start new Electronic Business companies. Customers or maybe suppliers will become competitors.

Harvard Business School (HBS) graduates and others specialize now in "attack plans." They take an existing business and develop a plan to attack it using Electronic Business. This attack strategy can be very successful. Witness Microsoft!

Assume that at least one company has each part of your business in its sights with a well-funded attack plan. Act accordingly, now!

Electronic Business will introduce different buying points for business. Some sectors will suffer—badly. Some will gain. Companies that can offer the best quality/price delivery will win. Delivery will include support as well as the product or service itself.

There will be a new competitive environment. Secondary players will be eliminated. This does not necessarily mean that the largest will survive. Size may be a huge obstacle to change and may open up markets to new competition. But Electronic Business does demand specialization. Companies must decide where they will attack, where they will defend and where they will run.

Imperative 7: Get Key Technology Players on Your Team

The seventh imperative is to lock up key technology players on your team if you can, so that at least their potential influence in your market is neutralized. Obviously, a technology company does not want to lose the potential to deal with all companies in a particular opportunity area unless the deal they are offered is so attractive they cannot refuse. There will thus be new methods of working with technology suppliers, many of which involve some form of partnership.

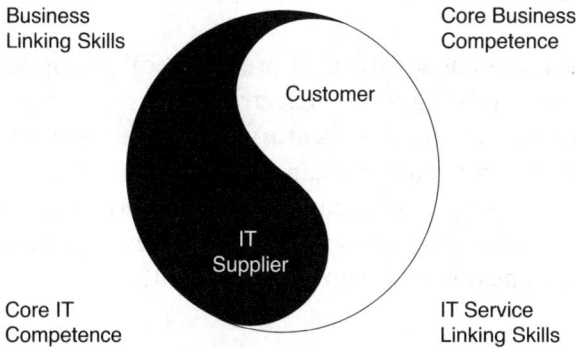

Figure 3. Creating a better fit for Electronic Business

Imperative 8: Buy When You Can; Sell When You Have To

Acquisition will thrive in this Electronic Business world. Companies are jockeying for position. This is like the early days in any industry change, whether it is airplanes, cars, telephones or computers. There are lots of opportunities and lots of players, as well as lots of money! So we will see a great deal of acquisition. Much of it will be at prices that do not seem to make sense. Tellabs in the USA, for example, paid $575 million for NetCore Systems Inc., an 85-employee company with no revenues in 1998!

One question must be: Will Electronic Business cause great consolidation in industries? Or will it lead to fragmentation? The answer is probably, both. But the concentration in any business will probably be far more transient than in the past, with the leading players changing frequently. The middle will be in constant flux, with companies moving up and some moving down.

When you have to have strength, buy. When a unit does not fit or cannot easily make the grade in the required time scale, sell it. Use leverage from financial resources such as venture capitalists (VCs) or investment funds to amplify your resources. If your company is valued at a price/earnings ratio of 10 and you have a substantial Electronic Business unit that can command a ratio of infinity, then use the public market for advantage.

Conclusion

The Web is a two-edged sword. It presents both an opportunity and a threat. It cannot be ignored. Electronic Business will be the way business is done in the 21st century. Our lives will be radically affected by it in some ways we can predict and in others we can only guess at. One thing is certain: Decisions made now will alter the course of company and national success, forever. The time to act in Electronic Business is now. Seize the moment!

Appendix A   Siemens Business Services

Faced with the challenge of the Electronic Business revolution, business leaders may ask themselves how they can best manage the transformation of their companies. "Do I have the resources on board? Do I have enough Electronic Business know-how to do it myself? Do I need outside help? If so, where do I find partners competent enough to guide me into new global business dimensions?" The answers to all of these questions will determine whether companies will be able to navigate the next industrial revolution, which is already under way now.

Siemens Business Services (SBS) has made it its mission to partner with companies and organizations to design, build and operate seamless Electronic Business solutions that both enhance and support the success of its customers. Access to best-in-class resources through a network-based organization and knowledge-sharing culture ensure the best solutions and services worldwide.

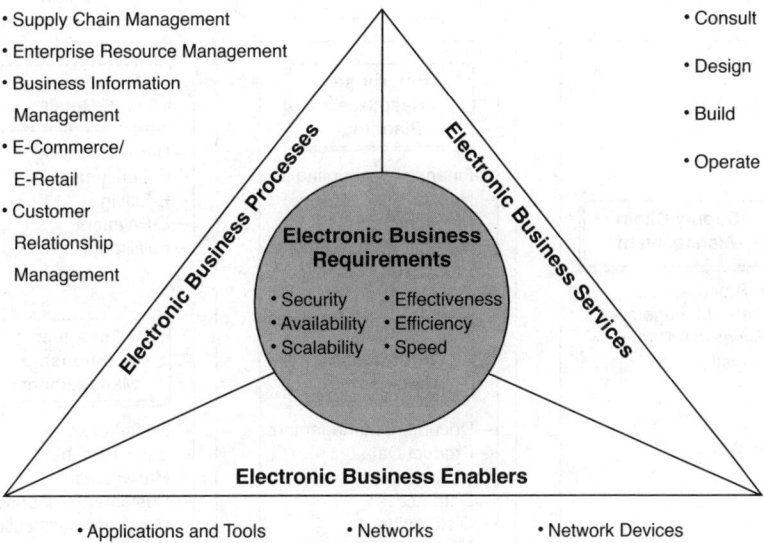

- Supply Chain Management
- Enterprise Resource Management
- Business Information
  Management
- E-Commerce/
  E-Retail
- Customer
  Relationship
  Management

Electronic Business Processes

Electronic Business Services

Electronic Business Requirements

- Security
- Availability
- Scalability
- Effectiveness
- Efficiency
- Speed

- Consult
- Design
- Build
- Operate

Electronic Business Enablers

- Applications and Tools
- Networks
- Network Devices

Figure 1. Electronic Business pyramid

Positioned for the Future

The three pillars of the SBS strategy are Business Integration, Business Management and Business Transformation.

For customers who require a business integration partner, SBS offers enterprise resource planning solutions; for clients in need of a business management partner, SBS has outsourcing and managed services solutions. Finally, when business changes demand a business transformation partner, SBS boasts a complete portfolio of digital business solutions, which can be described in an Electronic Business pyramid as shown in Figure 1.

At the center of the pyramid are the basic requirements for Electronic Business. They are security, availability, scalability, effectiveness, efficiency and speed. The foundation of all SBS Electronic Business solutions, they encompass the establishment of security standards for Electronic Business, the provision of 99.9 percent availability, the capability to scale up or down on a daily basis, strict price/performance evaluation, and adherence to corporate Electronic Business targets.

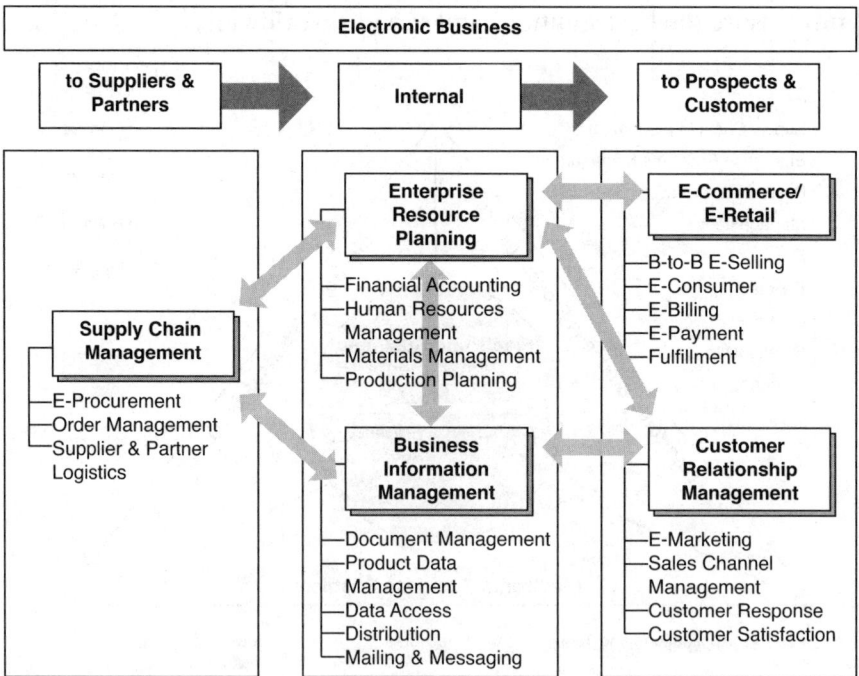

Figure 2. Electronic Business processes within the customer organization

The left side of the pyramid represents the core processes within a Electronic Business organization. Five "communities of practice" that apply across all industrial and market segments determine the future growth of organizations. They are Supply Chain Management, Enterprise Resource Management, Business Information Management, E-Commerce/E-Retail and Customer Relationship Management. Figure 2 shows how these practice communities relate to the value chain extending from suppliers and partners on the left through to customers and prospective customers on the right.

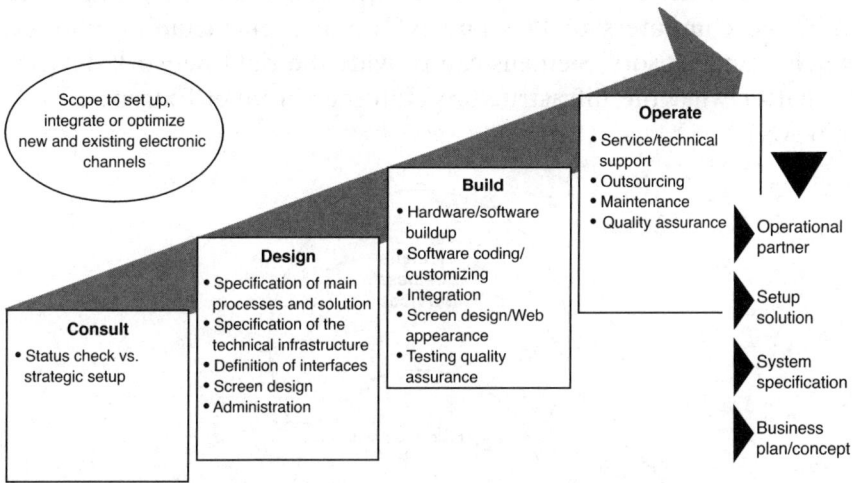

Figure 3. The complete range of SBS' Electronic Business services

The right side of the pyramid and Figure 3 show the services required to establish, integrate or optimize effective Electronic Business solutions. First, it is necessary to assess existing systems and solutions in light of the strategic goals; this is the consulting phase. For all service packages, the starting point is the definition of the processes, interfaces and the specification of the needed infrastructure; this is the design phase. The next step covers setting up all hardware and software components, coding and customizing the software, integrating new systems into the legacy environment, performing screen design, and testing the solution for quality; this is the building phase. The final phase is the operating phase. This involves the provision of service and technical support, outsourcing, maintenance, and constant quality assurance. For companies that want to concentrate on their core business and profit from leading-edge business processes, SBS specializes in business process outsourcing.

Built-In Convergence

Electronic Business enablers form the third important dimension. These are the applications and tools, the networks and network devices that finally make Electronic Business really work. This is where the combined strength of Siemens Information and Communications comes into play. This organizational structure puts Siemens in a position to drive Information and Communications convergence ahead. It also gives SBS access to the hardware, software, network integration and services needed for Electronic Business. From mainframe and midrange computers to PCs and NCs and even including phones, switches and sensors, Siemens can provide the right network devices, no matter what the infrastructure (Internet/Intranet/Extranet, LAN or WAN).

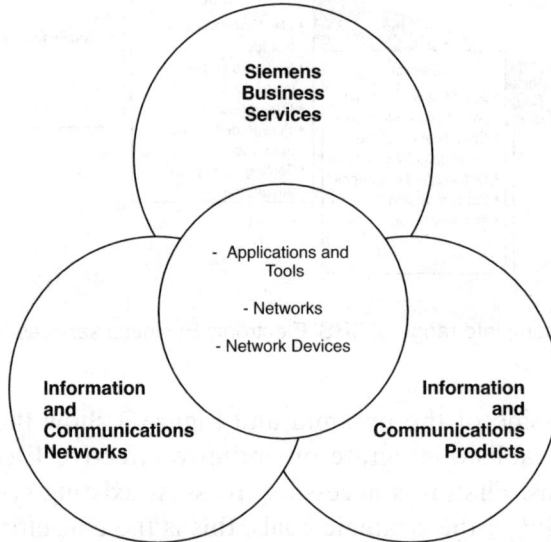

Figure 4. By combining its portfolio of information and communications products, networks and services, Siemens is uniquely able to exploit the synergies of I and C convergence

Moreover, SBS employs the best-in-class applications for Supply Chain Management (e.g., Manugistics, I2, SAP), Enterprise Resource Planning (e.g., SAP, PeopleSoft, Oracle, Kordoba), Business Information Management (e.g., Oracle, SAS, Lotus, Filenet), e-commerce (e.g., Netscape, BroadVision, Sterling Commerce) and Customer Relationship Management (e.g., Siebel, Clarify, Vantive).

Exemplary Electronic Business Solutions

To make Electronic Business a reality for customers around the world, SBS has developed a comprehensive set of service packages designed to help customers radically transform their business, products, services and operational effectiveness. These service packages are collectively called *e-SPEED* because the speed of transformation is mission-critical in the constantly changing Electronic Business marketplace. The modular approach of *e-SPEED* means that future expansion and access to new marketplaces are built into the SBS solution. Further service packages can be added as necessary. This guarantees fast response to new customer and business demands, giving clients the speed to stay ahead of the competition and position themselves as market leaders.

Siemens Business Services has already implemented a series of leading Electronic Business solutions for innovative customers across several industries. Listed below are some examples of Electronic Business at work.

Internet Service Provider for Garanti Bank

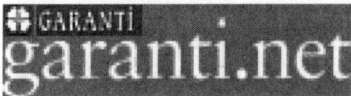

Challenge

Garanti Bank, recognizing that the World Wide Web would become a very important channel for banking services, made the decision to invest in Internet banking. To enlarge the Turkish Internet population, Siemens Business Services and Garanti Bank created garanti.net (www.garanti.net.tr), a new Internet Service Provider, to provide access to both the Internet backbone and Garanti Bank's Internet banking applications.

Solution

Businesses and private customers can now manage almost all of their banking needs online, as well as Internet browsing, e-mail and other value-added content provided via garanti.net. Garanti Bank has assumed responsibility for sales and marketing, using their call center, Internet site and in excess of 200 branches as sales channels for garanti.net.

Siemens Business Services is the technology provider and operator of the virtual Internet Service, establishing a fully digital backbone to meet the reliability and performance requirements. ISDN modems, latest-generation access servers, best-in-class mail server hardware and software, three-way satellite access to the Internet backbone, and five points of presence (Istanbul, European and Asian sides, Ankara, Izmir, and Bursa) in Turkey guarantee the quality standards of garanti.net. Siemens Business Services is responsible for the Internet backbone access, communication, IT infrastructure and system administration. The quality and efficiency of the technical operations, the subscriber management, help-desk functions, security, content engineering, and hosting of garanti.net are also the responsibilities of Siemens Business Services.

Within four months, garanti.net became the third-largest ISP in Turkey. The targeted number of users for the new service is 100,000 within three years.

Benefits

For Garanti Bank:

- Rapid entry into the Internet Service Provider market
- Transfer of economic risks to Siemens Business Services
- Focus remains on enhancing value added services around its core competencies—banking

... and for the Garanti Bank customers:

- Fastest Internet access via the ISDN infrastructure
- High quality ISP with subsidized prices
- The flexibility of Internet banking, including a 24-hour call center and help-desk support
- Access across all of Turkey

NetBank AG: An Internet-Only Bank

Challenge

NetBank AG, a company belonging to Sparda Banken, is Germany's first Internet-only bank. It offers services that go beyond the usual financial services and enable customers to get the most out of the possibilities of Internet banking. These include personalized information; account information accessible at any time; individually compiled, up-to-the-minute news; and electronic shopping for videos and CDs. Banking today should be reasonably priced, convenient, secure and individually tailored.

Solution

On behalf of and in cooperation with NetBank AG, Siemens Business Services implemented the entire one-to-one banking system. It also handled the integration of external information providers and of the module for payment transactions.

NetBank's Internet presence has been implemented on the basis of the One-to-One product from the company Broadvision. The information that visitors to the Web site provide about themselves is used to create profiles and dynamically generate offers tailored precisely to their requirements.

The more information the Web site has about a particular visitor, the greater the extent to which the Web pages and offers of products and services can be personalized (personalized marketing). By means of a three-level offer of information and services, the relationship to the Web site visitor is built up progressively.

Depending on the user profile, the following services are available:

- Anonymous visitor: general information about the bank, products and prices, press releases and financial information, a systematic search function, and access to services such as the online ordering of CDs

- Registered user: individual news profile, specially tailored securities information
- Bank customer: payment transactions, personalized banking

NetBank offers all standard transactions in the areas of payments and securities trading (from autumn '99). In addition to its financial services, NetBank provides securities information, a business dictionary, news and the opportunity to order CDs and videos.

Benefits:

For NetBank AG:

- Cultivation of attractive new customer segments
- Intensive customer relationship through one-to-one marketing
- Image gain as an innovator and technology leader
- Low marketing costs and lean organization

... and for its customers:

- High degree of flexibility
- Personalized, convenient service around the clock
- Attractively priced standard transactions
- Added value by means of non-bank offers

Bank Austria: One-to-One Banking

Challenge

The Bank Austria Group is the largest financial organization in Austria and is one of the top 40 banks in Europe. Despite this leading position, Bank Austria is aware that future competitiveness will depend on how well it can accommodate changing customer demands with new and innovative products and services. The integration of the bank's sales delivery channels is a key element of that strategy.

Solution

A personal home page for all. A new concept developed jointly with partners such as Siemens and BroadVision. In developing its home page, Bank Austria has embarked on an innovative design that is unique in Austria.

Technological progress over the last few years has made it possible to present each visitor with a personal home page. This ensures that not only personalized service and product offerings can be offered to users, but also that the users themselves can design the Bank Austria home page to suit their individual preferences and interests.

In addition to personalization of the home page, particular attention was also focused on simple user guidance, high operating speed, rapid access to online accounts, a completely re-designed e-mail service function and a search engine arranged alphabetically and according to topic. For the first time in the Austrian banking market, "communities of interest" with special chat and topic forums are being offered.

Benefits

* The first one-to-one marketing Internet banking concept in Austria (and one of the first in Europe) that will serve to enhance internal efficiency as well as improve customer service. Bank Austria is addressing an attractive customer segment.
* Open technology can be integrated into other electronic channels
* Financial and non-financial information specific to each community of interest
* Personalized information

TECCOM: E-Commerce for Vehicle Spares

TECCOM

Challenge

The vehicle spares industry and dealers initiated the TECCOM project to increase the competitiveness of the open vehicle spares business. The objective was to provide participants in the open market with a scalable system of electronic commerce. Using state-of-the-art technologies such as the Internet, online ordering and delivery processes between business partners and within companies had to be designed.

Solution

Developed by TECDOC, a joint venture of 28 parts manufacturers, and Siemens Business Services, TECCOM is a standardized procedure for electronic commerce that makes partner-specific applications superfluous. Companies involved in the business process and using this in company service include workshops, dealers and manufacturers.

From inquiry to order and confirmation of delivery through to billing, all business processes between the parts manufacturer, dealer and vehicle workshop can be handled electronically. Based on international standards for technology and data formats, trade solutions and system landscapes of different manufacturers can be linked to one another.

Benefits

For the workshop:

- Dramatic reduction in the use of phone and fax thanks to the electronic inquiry and ordering function
- Fewer transmission errors
- Cost transparency through access to electronic catalogs of various manufacturers

... and for dealers:

- Fewer mistakes in deliveries, thereby cutting warehouse movements and posting procedures, leading to less capital spending
- Individual design of information transmission and customer support

Siemens Mall: E-Commerce—Business to Business

Challenge

Formation of an electronic trading community that profits from being part of Siemens and from presenting a single, uniform face to the market. Developing the Siemens Mall as the Internet marketplace for Siemens companies to sell their products and services to business customers.

Within the community, the individual Siemens Groups can act nationally and internationally as independent service providers and at the same time plug their operational business processes into the overall system.

Solution

The Siemens Mall presents Siemens products to online purchasers in the form of catalogs compiled by Siemens Groups. Selecting products in the catalogs is easy thanks to a parameterized search function. The products are listed hierarchically and described using multimedia-style information that purchasers can call up as required.

They can use a dynamic query mechanism to make sure that what they want is currently available. The products and services selected can then be put into an electronic shopping cart, which can be stored as a specimen purchase so that future routine orders can be processed quickly and conveniently. Invoice addresses and delivery addresses are handled just as easily. Purchasers can store them for subsequent selection as new orders arrive. The payment procedures offered can likewise be selected on an order-specific basis or provided in the form of fixed links.

Prices are calculated on the basis of the quantity of products, with individual customer discounts being taken into account. At the click of a mouse, the system generates the order data from the shopping cart and sends it to the materials management system in the relevant Siemens Group. Messages on the status of the order tell purchasers what stage of processing their orders have reached at any particular time.

Benefits

For the Siemens Group:

- Streamlining of business processes in worldwide sales
- Group-specific product presentation
- "Community potential" thanks to the uniform international sales image (one face to the customer)

... and for Siemens business customers:

- Streamlining of business processes in procurement
- Individualized offers
- Up-to-date cross-Group Siemens offerings
- Convenient facilities for placing orders around the clock and any-where in the world

Smart Card Solution for Ruhr University Bochum

Challenge

Tomorrow's university will no longer be the traditional think tank of the 19th century, but a modern service company that offers custom-ers a product called "knowledge." The Internet and e-mail communi-cation, secure and fast network access, and electronic cash are all buzzwords that modern universities can no longer afford to ignore.

Solution

Three years ago, the Ruhr University in Bochum, Germany, launched a plan to offer its students a smart card with a variety of functions. Since then, some 10,000 "UniCards" have been issued. They can be used as student ID cards enabling the carriers to provide proof of identity, sign on at self-service terminals and make non-cash payments. Since 1998, Siemens has been advising the university with regard to finding new avenues of use and new target groups for its UniCards.

Starting in summer 1999, all students will be offered UniCards with extended cryptographic functions such as digital signatures. This helps lay the foundations for a virtual university that will provide students with access to global information sources and the chance to listen in to European lectures and take part in European examinations with a high level of security. Every new term sees an influx of some 6,000 new students; existing cards will be replaced successively.

Benefits

For the Ruhr University in Bochum:

- Optimized processes, particularly administration processes and payment transactions
- Forgery-proof identification thanks to optical and electronic tagging

... and for the students:

- Home workstations
- Much faster turnaround times and shorter wait times
- Secure access to services in the Intranet (e-mail, document loans)
- Payment with the UniCard (local public transport, e-commerce)
- Opportunity to study at a modern university, the universal virtual university

SBS Track Record

Formed in 1995, SBS has grown rapidly to become the No. 1 full-service provider in Germany, Europe's largest IT market, and one of the top four throughout Europe. Its sights are set high: In the corporate arena, SBS is aggressively pursuing its goal to become one of the world's top five full-service companies.

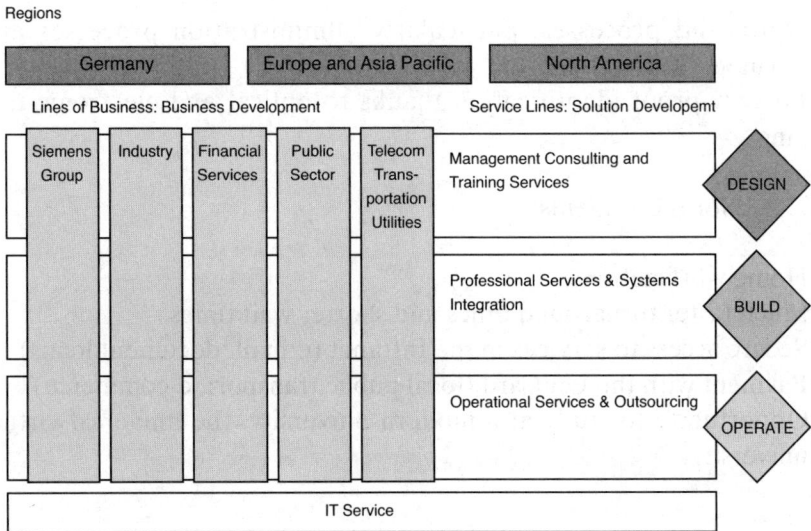

Regions

| Germany | Europe and Asia Pacific | North America |

Lines of Business: Business Development   Service Lines: Solution Developemt

Siemens Group	Industry	Financial Services	Public Sector	Telecom Transportation Utilities	Management Consulting and Training Services	DESIGN
					Professional Services & Systems Integration	BUILD
					Operational Services & Outsourcing	OPERATE

IT Service

Figure 5. Company structure

As a separate legal entity, SBS can act with great flexibility and agility while leveraging the global corporate resources of Siemens AG. One of three groups that constitute Siemens' Information and Communications business, SBS has assembled a global team of more than 21,000 professionals in over 50 countries that is second to none in Electronic Business expertise. Together, the Information and Communications Groups represent sales of DM 50 billion and form a powerful network poised to provide comprehensive, seamless Electronic Business solutions.

In order to best serve the unique requirements of each market sector, SBS has organized itself into the following lines of business: industry, financial services, public sector, telecommunications, transportation and utilities. So, the SBS professionals understand the business needs of each industry, making them uniquely able to identify and optimize the resources of their clients. Once identified, SBS harnesses business systems and technology to work together for maximum advantage through the use of leading-edge customer relationship management systems, innovative revenue-generating techniques, and the best IT and communications infrastructures.

Whatever the particular needs of client companies, SBS has the Electronic Business specialists to help them to be a winner in the Electronic Business revolution. SBS expertise spans a wide spectrum of services, which include enterprise resource planning, data warehousing, data mining, supply chain management and enterprise business development. By enlisting best-in-class vendors as partners, SBS guarantees its clients industry-leading solutions at the lowest possible total cost of ownership. All of SBS' solutions are offered along the design, build and operate continuum.

Table 1. SBS at a glance

Established	October 1995
CEO	Dr. Friedrich Fröschl
CFO	Michael Kutschenreuther
Revenues (97/98)	
World	Euro 3.17 billion (DM 6.2 billion)
(Germany)	Euro 1.89 billion (DM 3.7 billion)
Revenues by market sector	
Industry (including Siemens Group)	51 %
Financial Services	9 %
Public Sector	25 %
Telecom, Transportation, Utilities	15 %
Revenues by services provided	
Professional services and systems integration	59 %
Operational services and outsourcing	35 %
Consulting	3 %
Training	3 %
Employees (12/98)	
World	20,845
Germany	11,073
International presence	in 40 countries
Affiliates	34
SBS revenues compared to competition	
Germany	No. 1
Europe	No. 4
World	No. 9

Appendix B　　Glossary of terms

- **ADSL**, or asynchronous digital subscriber line: A variant of *DSL*.
- **AI**, or after Internet: some thing or event that has occurred after the dawn of the global Internet. This acronym can also stand for artificial intelligence (computer systems that mimic human intelligence).
- **ASP**, or application services provider: a company or organization developing and/or supplying computer systems applications·
- **ASR**, or automatic speech recognition: a computer system that recognizes human speech and utterances
- **ATM**, or automated teller machine: a bank system that replaces the functions of a human bank teller
- **Agent**: software that simulates intelligent or purposeful activities on behalf of a human user
- **Analog**: a thing that represents, is similar to or corresponds to another thing. In computing, usually the obverse of digital.
- **Application**: a computer program that accomplishes a purpose or set of purposes or provides functionality, e.g., a word processing *application*
- **Architecture**: the style or plan of a given system
- **Authentication**: in security systems, the function of verifying identity

- **B-to-B**, or business-to-business
- **B-to-C**, or business-to-consumer
- **Bandwidth**: the amount of carrying capacity of a communications line
- **Banner ad**: an advertisement posted, banner-style, on an Internet Web site
- **BI**, or before Internet: some thing or event that has occurred before the dawn of the global Internet
- **Broadband**: high-bandwidth communications systems, e.g., cable modem service, DSL service, etc.
- **Brochureware**: promotional material posted on an Internet Web site in exactly the same form in which it was prepared for print

- **Browser**: a software program enabling navigation and surfing of the Internet and the World Wide Web, e.g., Mosaic, Netscape Navigator, Microsoft Internet Explorer

- **C-to-B**, or consumer-to-business
- **CC&B**, or customer care and billing
- **CD**, or compact disc
- **CERN**: originally named "Conseil Europeen pour la Recherche Nucleaire," CERN was later renamed "European Laboratory for Particle Physics." The World Wide Web was invented here.
- **CPM**, or cost per thousand
- **C/S**, or client/server: a system of networked computers in which individual (or personal) computers are linked together and served software, applications and other services by a larger computer, or *server*
- **Cable modem**: a device that links a computer to a cable-delivered system providing access to broadband content
- **Channels**: stations in a network such as the Internet, analogous to TV channels
- **Chat (group)**: an Internet-enabled facility that allows people to exchange text or voice messages with others on the same service
- **Click-through**: the act of selecting an Internet posted advertisement by clicking via the browser on the part of the ad that is linked to another location (that of the advertiser) on the Internet
- **Client**: an individual computer that is linked together with other *client*s in a client/server network. See also *C/S* and *server*.
- **Code, coded**: the language of computer programs. Also, data that has been protected by an encryption system.
- **Collaboration**: using the Internet to enable a number of different people to work together, even from disparate geographical locations
- **Community**: a group with common bonds or interests, especially one enabled or facilitated by the Internet
- **Cookies**: software programs that track and record Web site visits
- **Cryptography**: a system for encoding data that can disguise its contents from unintended recipients
- **Cyberspace**: the virtual, electronic environment created by computer and telecommunications systems, especially by the Internet

- **DARPA**, or (US) Defense Advanced Research Projects Agency: a government organization famous for its research and development programs. It sponsored the original research that led to the development of, for example, the Internet.
- **DOS**, or disk operating system: The original operating system for early personal computers
- **DSL**, or digital subscriber line: A means of using installed copper wires to transmit broadband content
- **DUV**, or data under voice: a system for carrying both data and voice communications simultaneously over the same communications line. See also *VUD*.
- **Data center**: a central location for data processing and computing operations
- **Data visualization**: a means of displaying and simulating data relationships by visual and graphic means and methods
- **Database**: an organized and systematized way of arranging data sets
- **Desktop**: operations carried out upon or by using a computer, e.g., desktop publishing
- **Digital**: data represented and/or related to calculations with digits as opposed to physical representation (*analog*). Digital usually refers to representing switched states (on/off, yes/no, open/closed, etc.) by using the numerals zero (0) and one (1).
- **Disintermediation**: the process of removing intermediate layers. In the context of the Internet, usually refers to the elimination of jobs or services provided by middle-layer functions, such as distributors or jobbers, because the Internet allows for direct contact between makers and purchasers.
- **Distributed**: in the context of the Internet, the ability to spread out or widely distribute computing and communications functions and applications
- **Domain name**: an Internet address or *URL*

- **E1**: two megabit per second telecommunications line (the European equivalent of the T1 line in the United States)
- **EAS**, or enterprise applications solutions: application programs and systems designed for use throughout an entire business enterprise
- **EBPP**, or electronic bill payment and presentment
- **EC**: used as an acronym for both e-commerce (i.e., electronic or Internet-enabled commerce) and for European Community

- **EDI**, or electronic data interchange: a means of using a proprietary system for electronically exchanging data
- **ERP**, or enterprise resource planning: software systems designed for use throughout an entire business enterprise
- **E-commerce**, or electronic commerce
- **Electronic wallet**: a computer-based system for exchanging payments
- **E-mail**, or electronic mail: text communications sent via the Internet and stored for retrieval at the recipient's discretion
- **Embedded**: contained within an existing system, e.g., the system is *embedded* in the computer chip architecture
- **Encryption**: to encode data for security and secrecy. See also *cryptography*.
- **Explorer**: the brand name of the Microsoft Web browser software package
- **Extensible markup language**: see *XML*
- **Extranet**: a proprietary Internet-based network linking parts of a company's internal organization with outsiders (e.g., suppliers, customers, etc.)
- **Eyeballs**: audiences, i.e., visitors and users of a Web site

- **Fiber**: (also fibre) a kind of optical cabling used in communications and computing network systems. It uses light signals to transmit data at high speeds and with high capacity.

- **GDS**, or global distribution system
- **GDP**, or gross domestic product
- **GNP**, or gross national product
- **GPS**, or global positioning satellite
- **GUI**, or graphical user interface: an interface created using graphical elements. See also *interface*.
- **Gigabit**: a billion bits of data
- **Gigabyte**: a billion bytes of data
- **Groupware**: software systems designed to be used by multiple users in a collaborative manner

- **HDTV**, or high-definition television
- **HTML**, or hypertext markup language: the main language or coding system for representing data and graphic materials on the Internet and the World Wide Web

- **HTTP**, or hypertext transfer protocol: the language or coding system describing how Internet content can be located and accessed
- **Help desk**: in computing operations, the customer service facilities
- **High touch**: high technology systems that incorporate and integrate tactile and/or manipulative functions
- **Host**, and **hosting**: maintaining and storing the data that make up an Internet location, a Web site
- **Hub**: a Web site where various interrelated content is aggregated and combined. Similar to a *portal*.
- **Hyperlink**: a means of using the Internet to connect one data set or location to another via an electronic computer and communication system–mediated linkage

- **IAD**, or Internet access device(s)
- **ICT**, or information and communication technology: the common term used in Europe for the convergence of these two fields
- **IP**, or Internet protocol: the main communications method of encoding data packets for transport via the Internet
- **IPO**, or initial public offering (of stock shares)
- **ISP**, or Internet service provider: a business that provides access to the Internet
- **IT**, or information technology
- **Infrastructure**: the installed physical plant of equipment and systems
- **Integration**: the bringing together of disparate systems
- **Interface**: the system enabling function and usage. E.g., a computer mouse provides a point-and-click *interface* between a user and a computer system. See also *GUI*.
- **Internet**: a network of networks enabling distributed computing and communications on a global basis across many kinds and types of equipment and systems
- **Intranet**: an internal communications and computing system based on the Internet

- **Java**: an open computer programming language developed by Sun Microsystems. Programs written in Java (called Applets) can be used by different and otherwise incompatible computing systems

- **Kilobit**: about one thousand bits of data
- **Kilobyte**: about one thousand bytes of data

- **Kiosk**: a small structure, or virtual system, designed for facilitating content delivery or other types of transactions

- **Language**: a kind of code used to create computer instructions and programs
- **Laptop**: a small, portable computer system designed to be operated away from the desktop, perhaps on a person's lap. Often referred to as a *notebook*.
- **Legacy**: in computing systems, existing facilities
- **Linux**: an operating system based on the *UNIX* operating system. Famous for being developed by collaboration of computer programmers and systems analysts in what came to be known as an *open system*.

- **MP3**: a data compression standard that became famous as a way of coding audio music files so that they could be sent easily over the Internet
- **Macro**: a computer program designed to provide instructions for a specific and often repetitive function
- **Mainframe**: a very large computing system. Sometimes used as a *server* computer.
- **Markup (language)**: a coding system. See also *HTML*.
- **Megabit**: a million bits of data
- **Megabyte**: a million bytes of data
- **Messaging**: sending communications over data systems
- **Microcomputer**: a small computer, usually a *PC*
- **Microprocessor**: a chip containing *architecture* and systems for running a computer. Often called the "brains" of a computer.
- **Minicomputer**: a medium-sized computer. Sometimes used as a *server*.
- **Modem**, or modulator-demodulator: a device or software-based system designed to take computer output in digital form, modulate (convert) it into analog form to flow over telephony networks and modulate back again into digital form
- **Mosaic**: the first Internet *browser* system
- **Multimedia**: comprising or using multiple kinds of media

- **NC**, or network computer: designed for use over the Internet but not necessarily for use outside of Internet connectivity
- **Narrowcasting**: content designed for smaller audiences. Contrasted with broadcasting.

- **Navigator**: the brand name of the Netscape (now owned by America Online [AOL]) Web browser software package
- **Net**: a network, usually used to refer to the Internet
- **Network**: a system based upon communications lines that connect or interlink
- **Newsgroups**: on the Internet, a series of text-based discussion forums dealing with varied but specific content
- **Node**: a location or specific channel in a *network*
- **Notebook**: see *laptop*

- **Object**: a specific software element or programming component
- **Online**: of or occurring on the Internet or other electronic network
- **Online trading**: buying and selling of stocks, bonds and other financial instruments by means of a network, usually the Internet
- **Open systems**: software programs, operating systems and the like that were developed by collaborative means and are not proprietary to any one specific brand or system
- **Outsourcing**: hiring providers and workers outside of a company or organization in order to perform functions that might usually be done by internal employees

- **PC**, or personal computer
- **PDA**, or personal digital assistant: a computing and communications appliance, usually small enough to hold in one hand, that provides varied functions
- **PFS**, or personal financial software
- **POPs**, or points of presence: specific nodes on the Internet
- **POS**, or point of sale (system): a system designed to engage a customer at the point, place and time of a transaction
- **PTT**, or post telephone and telegraph: the legacy agency for handling communications. Often a governmental agency, usually used as a term outside the United States.
- **Packet, packet-switching**: Internet communications that work via the transfer and routing of packages of data (*packets*), which are often routed (*switched*) to and from various locations or nodes in the network
- **Pageviews**: the number of times a user looks at a specific page on an Internet Web site
- **Personalization**: the act of making and/or designing a system to meet requirements and desires of specific users
- **Pico**: a trillionth, or 10^{-12}

- **Platform**: a specific type or style of system, e.g., *platform-*independent
- **Portal**: a Web site designed to aggregate data and information and, thus, become a starting place for Web site surfing and navigation. Usually based around a *search* function, e.g., yahoo.com, etc.
- **Processor**: the component in a computer or other device that executes the instructions, i.e., processes the data
- **Protocol**: a formal code

- **SKUs**, or stock keeping units
- **SOHO**, or small office/home office: used as a description of a specific demographic marketplace
- **Scale**: the ability to transition from one situation to another, especially with respect to size or breadth. E.g., the computer system could *scale* easily from client/server all the way to the *Internet*.
- **Search**, **search engine**: a facility for looking for and finding data types across the Internet or parts of it
- **Searchable archive**: a collection of data and information that is amenable to a search facility
- **Server**: a computer system used to link up with other *client* computers
- **Set-top box**: a device used to decode incoming data streams from, for example, cable TV systems
- **Site visits**: the number of times users surf to a specific Web site on the Internet
- **Smart cards**: credit card devices with embedded systems or software that add facilities or intelligence
- **Snail mail**: used to refer to mail carried via traditional post office or other physical carriers, as opposed to e-mail
- **Speech recognition**: a computer-based system that understands spoken words and utterances. Also known as *ASR*.
- **Standard**: an approach that has been accepted for use across an entire system. E.g., HTML is the *standard* for coding Internet content.
- **Supercomputer**: an extremely large computing system. This is a moving target. The power of supercomputers from the 1980s is now available on some PCs and desktop computers.
- **Surf**: to move about the Internet, visiting one Web site and then another, and another …

- **T1 & T3**: 1.5 megabit per second telecommunications lines in the United States (the US equivalent of the *E1* line in Europe)
- **TCP/IP**, or transmission command protocol/Internet protocol. The *standard* for transmitting data over the Internet.
- **Touchpad**: an *interface* device for directing the cursor on a computer screen. Usually used in place of a mouse.
- **Touchscreen**: an *interface* device that allows the user to move the cursor or otherwise direct other computing operations by touching various points on a computer's display screen
- **Tracking stock**: an equity stock used to isolate a specific set of corporate functions. E.g., GM-H stock is the *tracking stock* for the Hughes Electronics division of GM.
- **Traffic**: aggregated data communications
- **Turnkey**: a system that can be easily set up to begin operations without extensive preparations or instructions. E.g., one need merely "turn the key" in order to start the car.

- **URL**, or universal resource locater: a formal Internet assigned address or *domain name*
- **Universal access**: a system that has broad and diverse availability. E.g., the spread of the Internet has provided many with *universal access*.
- **UNIX**: a computer operating system originally developed at AT&T Bell Laboratories. See also *Linux*.

- **VAN**, or value-added network
- **VOI**, or voice over Internet: a telephony system that uses the Internet to send voice communications in packets
- **VPN**, or virtual private network: a network set up for a specific and limited set of functions and processes
- **VUD**, or voice under data: a system for carrying both data and voice communications simultaneously over the same communications line. See also *DUV*.
- **Video conferencing**: using computers and communications systems to transmit live videos of meetings and the like
- **Virtual**: existing in effect but not necessarily as a physical entity
- **Visualization**: used in the computing world to indicate the translation from data into a graphic or visualizable form
- **Vortal**, or vertical portal: a portal site for a focussed range of information, products and services

- **Web**, or World Wide Web, or WWW: a section of the Internet that allows the communication and transfer of rich data types such as graphics, images and video
- **Webcasting**: transmitting over the Web.
- **Web site**: a specific location where data and information are stored on the Web
- **Windows**: a way of dividing the space on a computer display into specific and separate areas or windows. Also, the brand name of a Microsoft operating system, e.g., Windows 98, Windows NT, etc.
- **Wireless**: without wires, i.e., communication that does not employ wires or cables to transmit data, voice, etc.
- **Workstation**: a type of computer, usually, though not exclusively, used to denote a more advanced version of a computer than a PC or desktop

- **XML**, or extensible markup language: an advanced Internet coding system, designed to work in conjunction with *HTML* and to add new information over and above that indicated in the HTML coding

- **Y2K**, or Year 2000 coding problem: Many, especially older, computer systems were made to recognize a year date by referring to two digits, e.g., 95 for 1995. In the year 2000, these systems will mistake 00 as the year 1900, thus causing a variety of errors and error-ridden computing.

Printing: Saladruck, Berlin
Binding: Buchbinderei Lüderitz & Bauer, Berlin